The Insecticide, Herbicide, Fungicide Quick Guide

by
B. G. PAGE
W. T. THOMSON

1995 Revision

THOMSON PUBLICATIONS
P. O. BOX 9335
FRESNO, CA 93791
(209) 435-2163
FAX (209) 435-8319

THIS BOOK FOR REFERENCE ONLY

READ THE LABEL CAREFULLY

INTRODUCTION

This book is designed to be used as an everyday tool for the person making pesticide recommendations. It can be used as a fast, efficient guide providing useful information as to which chemical should be used to control a certain pest on a specific crop. Information in this book was obtained from the manufacturers' labels and the EPA pesticide summary. There will also be a certain amount of variation due to state registrations and constant registration additions and deletions. A revision is made yearly to keep the book up-to-date. This book is designed as a guide and should be counter-checked by labels and local or state authorities. **READ THE LABEL.**

HOW TO USE THIS BOOK

This book is designed to be as helpful as possible to the reader. After using the book the publisher would appreciate any of your comments on making it more useful in the future revisions.

First, each group of pesticides are cross-referenced to the crop they are registered for use on. Then, by turning to the back of each pesticide section, the chemicals are listed showing which pests each one will control. By so doing you can make a general recommendation on which material to use in each individual situation.

At the end of the chemical section are listed the addresses of the basic manufacturer of the chemicals listed, as well as conversion and calibration charts.

TABLE OF CONTENTS

Page No.

INSECTICIDES

TRADE NAME CONVERSION TABLE — INSECTICIDES

Abate	*Temephos*
Acrobe	*B.t. var israelensis*
Actellic	*Pirimiphos-Methyl*
Admire	*Imidacloprid*
Affirm	*Avermectin B*
Agree	*B.t. var. aizawai*
Agri-Mek	*Avermectin B*
Align	*Azadirachtin*
Altosid	*Methoprene*
Ambush	*Permethrin*
Amdro	*Hydramethylnon*
Ammo	*Cypermethrin*
Apex	*Methoprene*
Apollo	*Clofentezine*
Armor	*Cyromazine*
Asana	*Esfenvalerate*
Astro	*Permethrin*
Atroban	*Permethrin*
Avid	*Avermectin B*
Award	*Fenoxycarb*
Azatin	*Azadirachtin*
Bactimos	*B.T. Var Isralensis*
Bactospeine	*Bacillus-thuringiensis*
Basamid	*Dazomet*
Baygon	*Propoxur*
Baythroid	*Cyfluthrin*
Benefit	*Azadirachtin*
Biflex	*Bifenthrin*
Biobit	*Bacillus-thuringiensis*
Bioneem	*Azadirachtin*
BioPath	*Metharhizium anisopillae*
Bolstar	*Sulprofos*
Bora-Care	*Inorganic Borate*
Brigade	*Bifenthrin*
Broot	*Trimethacarb*
Capture	*Bifenthrin*
Carzol	*Formetanate*
Checkmate	*Gossyplure*
Checkmate TPW	*Lycolure*
Citation	*Cyromazine*
Co-Ral	*Coumaphos*
Comite	*Propargite*
Commodore	*Lambda-cyhalothrin*
Condor	*B.T. strain 2348*
Conquer	*Esfenvalerate*
Counter	*Terbufos*

Crusade	*Fonophos*
Cryolite	*Sodium fluoaluminate*
Curacron	*Profenofos*
Cutlass	*B.t. strain EG 2371*
Cygon	*Dimethoate*
Cymbush	*Cypermethrin*
Cynoff	*Cypermethrin*
Cythion	*Malathion*
DDVP	*Dichlorovos*
Danitol	*Fenpropathrin*
Deadline	*Metaldehyde*
Decoy	*Gossyplure*
Decoy TPW	*Lycolure*
Demize	*Linalool*
Demon	*Cypermethrin*
Derringer	*Resmethrin*
Design	B.t. var aizawai
Di-Syston	*Disulfoton*
Diacon	*Methoprene*
Dianex	*Methoprene*
Dibrom	*Naled*
Dimilin	*Diflurenzuron*
Dipel	*Bacillus-thuringiensis*
Dipterex	*Trichlorfon*
Dursban	*Chlorpyrifos*
Dycarb	*Bendiocarb*
Dyfonate	*Fonofos*
Dylox	*Trichlorfon*
Eclipse	*Fenoxycarb*
Ectiban	*Permethrin*
Ectrin	*Fenvalerate*
Enstar	*Kinoprene*
Estate	*Chlorpyrifos*
Ficam	*Bendiocarb*
Fireban	*Tefluthrin*
Fluorguard	*Sulfluramid*
Foil	*B.t. strain EG 2424*
Force	*Tefluthrin*
Fumitoxin	*Aluminum Phosphide*
Funi-Cel	*Magnesium Phosphide*
Furadan	*Carbofuran*
Fury	*Zeta-cypermethrin*
Fyfanon	*Malathion*
Gencor	*Hydroprene*
Gentrol	*Hydroprene*
Grasshopper Spore	*Nosema Loctustae*
Guthion	*Azinophos-Methyl*
Imidan	*Phosmet*

Javelin	*Bacillus-thuringiensis*
Karate	*Lambda-Cyhalothrin*
Karathane	*Dinocap*
Kelthane	*Dicofol*
Knox-out	*Diazinon*
K-Othrine	*Deltamethrin*
Kryocide	*Sodium Fluoaluminate*
Lannate	*Methomyl*
Larvadex	*Cyromazine*
Larvin	*Thiodicarb*
Legion	*Naled*
Lepid	*B.t. var kurstaki*
Lock-On	*Chlorpyrifos*
Logic	*Fenoxycarb*
Lorsban	*Chlorpyrifos*
M-Pede	*Insecticidal Soap*
M-Peril	*B.t. var kurstaki*
M-Trak	*B.t. var tenebrionis*
MVP	*Bacillus thuringiensis*
MVP	*B.t. var kurstaki*
Magtoxin	*Magnesium Phosphide*
Mainstay	*Fonofos*
Marathon	*Imidacloprid*
Margosan-O	*Azadirachtin*
Marlate	*Methoxychlor*
Mavrik	*Fluvalinate*
Max Force	*Hydremethylnon*
MEC	*Gossyplure*
Merit	*Imidacloprid*
Minex	*Methoprene*
Mitac	*Amitraz*
Mocap	*Ethoprophos*
Monitor	*Methamidophos*
Morestan	*Oxythioquinox*
Mycotrol	*Beauvaria bassiana*
NAF-46	*Hexaflumuron*
Naturalis-L	*Beauvaria bassiana*
Neemisis	*Azadirachtim*
Neemix	*Azadirachtin*
Nemacur	*Fenamiphos*
Nolo Bait	*Nosema Loctustae*
Nomate TPW	*Lycolure*
Nu-Gro	*Pirimiphos-methyl*
Oftanol	*Isofenphos*
Omite	*Propargite*
Ornamite	*Propargite*
Orthene	*Acephate*
Otto	*Acephate*

Ovasyn	*Amitraz*
PBW Rope	*Gossyplure*
Pageant	*Chlorpyrifos*
Penncap-M	*Methyl parathion*
Pentac	*Dienochlor*
Pestroy	*Fenitrothion*
Petcor	*Methoprene*
Pharorid	*Methoprene*
Phaser	*Endosulfan*
Phostoxin	*Aluminum Phosphide*
Plantfume	*Sulfotepp*
Pounce	*Permethrin*
Pramex	*Permethrin*
Precision	*Fenoxycarb*
Precor	*Methoprene*
Prelude	*Permethrin*
Provado	*Imidacloprid*
Proxol	*Trichlorfon*
Pryfon	*Isofenphos*
Pydrin	*Fenvalerate*
Pyrid	*Fenvalerate*
Rabon	*Tetrachlorvinphos*
Rampart	*Phorate*
Raptor	*B.t. var kurstaki*
Reldan	*Chlorpyrifos-Methyl*
Rotate	*Bendiocarb*
Ryana	*Rayanocide*
Safer	*Insecticidal Soap*
Safrotin	*Propetamphos*
Saf-T-Side	*Petroleum Oil*
Saga	*Tralomethrin*
Savey	*Hexythiazok*
SBP1382	*Resmethrin*
Scimitar	*Lambda-cyhalothrin*
Scourge	*Resmethrin*
Scout X-Tra	*Tralometrin*
Sectagon II	*Metham-NA*
Seige	*Hydroprene*
Sevin	*Carbaryl*
Skeetal	*B.T. Var Isralensis*
Soil Prep	*Metham-NA*
Spod-X	*Spodoptera exigua NPV*
Sumithrin	*Fenothrin*
Sun Spray	*Petroleum Oil*
Supracide	*Methidathion*
Suspend	*Deltamethrin*
Taktic	*Amitraz*
Talstar	*Bifenthrin*

Tame	*Fenpropathrin*
Teknar	*B.T. Var Isralensis*
Temik	*Aldicarb*
Tempo	*Cyflothrin*
Tenure	*Chlorpyrifos*
Thimet	*Phorate*
Thiodan	*Endosulfan*
Thuricide	*Bacillus-thuringiensis*
Timbor	*Inorganic Boron*
Torpedo	*Permethrin*
Torus	*Fenoxycarb*
Tribute	*Fenvalerate*
Trident	*B.T. Var Tenebrionis*
Trigard	*Cyromazine*
Triumph	*Isazofos*
Turcam	*Bendiocarb*
Turplex	*Aladirachtin*
Vapam	*Metham-NA*
Vapona	*Dichlorovos*
Vault	*B.t. var kurstaki*
Vectobac	*B.T. Var Isralensis*
Vectrin	*Resmethrin*
Vendex	*Fenbutatin-Oxide*
Veratran D	*Sabadilla*
Vikane	*Sulfuryl floride*
Volcano	*Sulfluramid*
Volck Oil	*Petroleum Oil*
Vydate	*Oxamyl*
Xentari	*B.t. var aizawai*
Yardex	*Fluvalinate*
Zephyr	*Avermectin B*

INSECTICIDES

AG PREMISES
Abate
Acrobe
Affirm
Altosid
Amdro
Asana
Astro
Avermectin
Azadirachtin
Azatin
Bacillus thuringiensis
Basamid
Baygon
Baytex
Biflex
BioPath
Boric Acid
BTI
Commodore
Condor
Cutlass
Cypermethrin
DDVP
Derringer
Demize
Diazinon
Dibrom (Naled)
Dimethoate
Dimilin
Dipterex
Dursban
Dylox
Ectiban
Esfenvalerate
Fenvalerate
Ficam-W
Furadan
Gencor
Gentrol
Guthion
Hexaflumuron
Inorganic borate

Karate
Kelthane
K-Othrine
L. giganteum
Lindane
Logic
Malathion
Margosan-O
Methaldehyde
Methyl bromide
Mycotrol
Naturalis-L
Nosema spore
Oftanol
Orthene
Permethrin
Pharorid
Precor
Pyrethrin
Rabon
Resmethrin
Rotenone
Safrotin
Seige
Sevin
Sulfluramid
Tempo
Tetralate
Tim-bor
Torus
Turcam
Vikane
Yardex

ALFALFA
Adios
Bacillus thuringensis
Condor
Cutlass
Dimethoate
Furadan
Guthion
Imidan

Lorsban
Malathion
Methomyl
Methoxychlor
Methaldehyde
Mycotrol
Parathion
Permethrin
Pyrellin
Pyrethrin
Rotenone
Sevin
Spod-X
Telone
Telone C-17
Thiodan
Xentari

ALMONDS
Agree
Apollo
Asana
Azadirachtin
Bacillus thuringiensis
Condor
Cutlass
Diazinon
Dibrom
Fumi-cel
Guthion
Imidan
Insecticidal Soap
Lorsban
Methaldehyde
Methyl bromide
MG phosphide
Omite
Permethrin
Petroleum
    Oil-dormant
    and summer oils
Phostoxin
Pyrellin

Pyrethrin
Sevin
Supracide
Telone
Telone C-17
Vendex
Xentari

ANISE
Align
Azadirachtin
Methomyl
Permethrin
Pyrellin

APPLES
Asana
Azadirachtin
Bacillus thuringiensis
Carzol
Codlure
Condor
Cutlass
Diazinon
Dibrom
Dimethoate
Di-Syston
Guthion
Imidan
Insecticidal soap
Kelthane
Lorsban
Malathion
Metaldehyde
Methomyl
Methoxychlor
Methyl bromide
Morestan
MPV
Nemacur
Omite
Permethrin
Petroleum
   Oil-dormant
   and summer oils

Phosphamidon
Provado
Pyrethrin
Rotenone
Ryanocide
Sevin
Supracide
Tame
Telone
Telone C-17
Thoidan
Vapam
Vendex
Vydate
Xentari

APRICOTS
Adios
Apollo
Asana
Azadirachtin
Bacillus thuringiens
Condor
Cryolite
Cutlass
Di-Syston
Diazinon
Dibrom
Imidan
Kelthane
Malathion
Methaldehyde
Methoxychlor
Methyl bromide
Morestan
Omite
Petroleum
   Oil-dormant
   and summer oils
Pyrellin
Pyrethrin
Sevin
Supracide
Telone
Telone C-17

Thiodan
Vapam

ARTICHOKES
APM Ropes
Asana
Azadirachtin
Bacillus thuringiensis
Biosafe
Condor
Cutlass
Furadan
Methaldehyde
Permethrin
Pyrellin
Pyrethrin
Supracide
Thiodan

ASPARAGUS
Azadirachtin
Bacillus thuringiensis
Biosafe
Condor
Cutlass
Disyston
Dyfonate
Insecticidal Soap
Lorsban
Malathion
Methaldehyde
Methomyl
Methoxychlor
Methyl bromide
Nemacur
Permethrin
Petroleum Oils
Pyrellin
Pyrethrin
Rotenone
Sevin
Spod-X
Telone
Telone C-17
Vapam
Xentari

10

AVOCADO

Align
Bacillus thuringiensis
Condor
Cutlass
Malathion
Methaldehyde
Methomyl
MG Phosphide
Permethrin
Petroleum Oils
Pyrellin
Sevin
Xentari

BANANAS

Align
Bacillus thuringiensis
Condor
Counter
Cutlass
Diazinon
Dyfonate
Dylox
Furadan
Lorsban
Methaldehyde
MG Phosphide
Mocap
Nemacur
Petroleum
  Oil-dormant
  and summer oils
Rugby
Sevin
Vydate
Xentari

BARLEY

Bacillus thuringiensis
Condor
Cutlass
Di-Syston
Fumi-cel
Furadan

Guthion
Lindane
Malathion
Methaldehyde
Methomyl
Methoprene
Methoxychlor
Methyl bromide
Methyl parathion
Mycotrol
Parathion
Phostoxin
Pyrethrin
Reldan
Sevin
Thiodan
Xentari

BEANS (all)

Adios
Agree
Asana
Azadirachtin
Bacillus thuringiensis
Biosafe
Comite
Condor
Cutlass
Dibrom
Dimethoate
Di-Syston
Dyfonate
Enstar
Insecticidal Soap
Kelthane
Lindane
Lorsban
Malathion
Methaldehyde
Methomyl
Methoxychlor
Methyl bromide
Mocap
Orthene
Petroleum Oils

Pyrellin
Pyrethrin
Sevin
Slam
Spod-X
Telone
Telone C-17
Temik
Thimet
Thiodan
Xentari

BEESWAX

Amitraz

BEETS

(Table or Red)
Adios
Azadirachtin
Bacillus thuringiensis
Biosafe
Condor
Cutlass
Diazinon
Dyfonate
Insecticidal Soap
Kryocide
Malathion
Methaldehyde
Methomyl
Methoxychlor
Methyl bromide
Nemacur
Petroleum Oils
Phosdrin
Proxol
Pyrellin
Pyrethrin
Sevin
Spod-X
Telone
Telone C-17
Thiodan
Xentari

BLUEBERRIES
Agree
Align
Bacillus thuringiensis
Condor
Cutlass
Diazinon
Guthion
Imidan
Insecticidal Soap
Kryocide
Lorsban
Malathion
Methaldehyde
Methomyl
Methoxychlor
MVP
Petroleum
  Oil-dormant
  and summer oils
Pyrellin
Pyrethrin
Sevin
Telone
Thiodan

BOK CHOY
Agree
Azadirachtin
Bacillus thuringiensis
Condor
Cutlass
MVP
Nemacur

BRAZEL NUT
Phostoxin
MG Phosphide

BROCCOLI
Agree
Asana
Azadirachtin
Bacillus thuringiensis
Biosafe
Condor
Cutlass

Diazinon
Dibrom
Dimethoate
Di-Syston
Dyfonate
Guthion
Insecticidal soap
Kryocide
Larvin
Lindane
Lorsban
Malathion
Methaldehyde
Methomyl
Methoxychlor
Methyl bromide
Monitor
MPV
Permethrin
Pyrellin
Pyrethrin
Sevin
Spod-X
Telone
Telone C-17
Thiodan
Vapam
Xentari

BRUSSELS SPROUTS
Agree
Azadirachtin
Bacillus thuringiensis
Biosafe
Condor
Cutlass
Diazinon
Dibrom
Dimethoate
Di-Syston
Dyfonate
Guthion
Insecticidal soap
Kryocide
Lindane

Lorsban
Methaldehyde
Methomyl
Methoxychlor
Methyl Bromide
Monitor
MPV
Nemacur
Nicotine
Orthene
Permethrin
Pyrellin
Pyrethrin
Sevin
Telone
Telone C-17
Thiodan
Vapam
Xentari

BUCKWHEAT
Malathion
Methoprene
Pyrethrins
Telone
Telone C-17

CABBAGE
Agree
Ammo
Asana
Azadirachtin
Bacillus thuringiensis
Biosafe
Condor
Cutlass
Diazinon
Dibrom
Dimethoate
Di-Syston
Dyfonate
Guthion
Insecticidal soap
Kryocide
Larvin

Lindane
Lorsban
Malathion
Methaldehyde
Methomyl
Methoxychlor
Methyl Bromide
Mocap
Monitor
MPV
Naturalis-L
Nemacur
Permethrin
Petroleum Oils
Pyrellin
Pyrethrin
Sevin
Spod-X
Telone
Telone C-17
Thiodan
Vapam
Xentari

CANEBERRIES
Agree
Align
Bacillus Thuringiensis
Condor
Cutlass
Cryolite
Diazinon
Furadan
Guthion
Insecticidal Soap
Kryocide
Lorsban
Malathion
Methaldehyde
Methoxychlor
Nemacur
Petroleum
 Oil-dormant
 and summer oils
Pyrellin

Pyrethrin
Sevin
Telone
Telone C-17
Thiodan
Vendex
Xentari

CANOLA
Cutlass
MPV
Xentari

CARABOLA
Pyrellin
Supracide

CARROTS
Asana
Azadirachtin
Bacillus thuringensies
Biosafe
Condor
Cutlass
Diazinon
Insecticidal Soap
Kryocide
Methaldehyde
Methomyl
Methoxychlor
Methyl Bromide
Methyl parathion
Pyrellin
Pyrethrin
Sevin
Telone
Telone C-17
Thiodan
Vydate
Xentari

CASHEW
Azadirachtin
Phostoxin
MG Phosphide

CATTLE
Chlorpyrifos
Cypermethrin
Co-ral
DDVP
Diazinon
Dibrom
Fenthion
Methoxychlor
Neguvon
Permethrin
Petroleum Oil
Pine Oil
Pyrethrin
Rabon
Taktic

CAULIFLOWER
Agree
Asana
Azadirachtin
Bacillus thuringiensis
Biosafe
Condor
Cutlass
Diazinon
Dibrom
Dimethoate
Di-Syston
Dyfonate
Guthion
Insecticidal soap
Kryocide
Larvin
Lindane
Lorsban
Methaldehyde
Methomyl
Methoxychlor
Methyl bromide
Monitor
MPV
Orthene
Permethrin
Petroleum Oils

Pyrellin
Pyrethrin
Sevin
Spod-X
Telone
Telone C-17
Thiodan
Vapam
Xentari

CELERY
Agree
Agri-mek
Azadirachtin
Bacillus thuringiensis
Biosafe
Clandosan
Condor
Cutlass
Diazinon
Dibrom
Dimethoate
Guthion
Insecticidal Soap
Larvin
Lindane
Malathion
Methaldehyde
Methomyl
Monitor
Orthene
Permethrin
Petroleum Oils
Pyrellin
Pyrethrins
Sevin
Spod-X
Telone
Telone C-17
Thiodan
Trigard
Vapam
Vydate
Xentari

CHAYOTE
Malathion

CHERIMOYA
Lorsban
Pyrellin

CHERRIES
Apollo
Asana
Azadirachtin
Bacillus thuringiensis
Condor
Cutlass
Diazinon
Dimethoate
Di-Syston
Guthion
Imidan
Insecticidal Soap
Kelthane
Lorsban
Malathion
Methaldehyde
Methyl bromide
Methoxychlor
Morestan
Nemacur
Omite
Petroleum
  Oil-dormant
  and summer oils
Pyrellin
Pyrethrin
Sevin
Supracide
Telone
Telone C-17
Thiodan
Vapam
Vendex
Xentari

CHESTNUTS
Bacillus thuringiensis
Condor

Cutlass
Diazinon
Insecticidal Soap
Kelthane
Malathion
Methyl bromide
Petroleum Oils
Pyrellin
Pyrethrin
Sulfur
Telone
Telone C-17
Thiodan

CHICORY
Cutlass

CHINESE CABBAGE
Agree
Asana
Azadirachtin
Bacillus thuringiensis
Biosafe
Condor
Cutlass
Insecticidal Soap
Lorsban
Methomyl
MG Phosphide
MPV
Permethrin
Pyrellin
Sevin
Trigard
Vapam
Xentari

CHINESE MUSTARD
Trigard

CHINESE RADISH
Agree
Diazinon

CIPOLLINI BULBS
Methyl Bromide

CITRON
Methyl Bromide
MG Phosphide
Petroleum Oils

CITRUS
Agri-mek
Azadirachtin
Bacillus thuringiensis
Biovector
Carzol
Comite
Condor
Counter
Cutlass
Dibrom
Dimethoate
Eclipse
Enzone
Ethion
Guthion
Insecticidal soap
Kelthane
Kryocide
Logic
Lorsban
Malathion
Methaldehyde
Methomyl
Methyl bromide
MG Phosphide
Mitac
Morestan
MPV
Nemacur
Nicotine
Omite
Petroleum
   Oil-dormant
   and summer oils
Pyrellin
Pyrethrin
Rotenone
Rotate
Ryanocide

Sabadilla
Sevin
Sulfur
Supracide
Telone
Telone C-17
Temik
Thiodan
Vendex
Vydate
Xentari

CLOVER
Adios
Bacillus thuringiensis
Comite
Lindane
Malathion
Methaldehyde
Methoxychlor
Pyrellin
Pyrethrin
Sevin
Telone
Telone C-17

COCOA
Methyl Bromide
MG Phosphide
Nemacur
Phostoxin
Pyrethrins

COFFEE
Align
Counter
Di-Syston
Furadan
Mavrik
Methyl Bromide
Mg Phosphide
Phostoxin
Temik
Thimet

COLLARDS
Agree
Asana
Azadirachtin
Bacillus thuringiensis
Biosafe
Condor
Cutlass
Diazinon
Dibrom
Insecticidal Soap
Lindane
Lorsban
Malathion
Methaldehyde
Methomyl
Methoxychlor
MPV
Permethrin
Pyrellin
Pyrethrin
Sevin
Telone
Telone C-17
Thiodan
Xentari

COPRA
Ethylene oxide
Pyrethrins

CORN
Actellic
Adios
Agree
Asana
Azadirachtin
Bacillus thuringiensis
Biosafe
Broot
Capture
Comite
Condor
Counter
Cutlass

Diacon
Diazinon
Dimethoate
Di-Syston
Dursban
Dyfonate
Force
Fumi-cel
Furadan
Insecticidal Soap
Kryocide
Larvin
Lindane
Lorsban
Malathion
Methaldehyde
Methomyl
Methoxychlor
Methyl bromide
Methyl parathion
Mg Phosphide
Mocap
M-Peril
MPV
Mycotrol
Parathion
Permethrin
Petroleum Oils
Phostoxin
Pyrellin
Pyrethrin
Reldan
Ryanocide
Sevin
Slam
Spod-X
Thimet
Thiodan
Vapam

COTTON
Admire
Asana
Bacillus thuringiensis
Baythroid

Bidrin
Bolstar
Capture
Comite
Condor
Curacron
Cutlass
Cypermethrin
Danitol
Design
Dibrom
Dimethoate
Dimilin
Di-Syston
Fumi-cel
Furadan
Fury
Gossyplure
Guthion
Karate
Kelthane
Larvin
Lindane
Lorsban
Malathion
Mavrik
Methomyl
Methyl bromide
Methyl parathion
MG Phoshide
Monitor
Mycotrol
Naturalis-L
Nemacur
Orthene
Ovasyn
Parathion
Phostoxin
Pyrellin
Pyrethrin
Scout
Sevin
Spod-X
Supracide
Telone

Telone C-17
Temik
Thimet
Thiodan
Vydate
Zephyr

COWPEAS
(Black-eyed beans,
Southern peas)
Adios
Bacillus thuringiensis
Di-Syston
Guthion
Insecticidal Soap
Lorsban
Methaldehyde
Methomyl
Methoxychlor
Pyrellin
Pyrethrin
Sevin
Telone
Telone C-17

CRABAPPLES
Align
Bacillus thuringiensis
Di-Syston
Guthion
Imidan
Kelthane
Petroleum Oil
Pyrellin

CRANBERRIES
Agree
Align
Biosafe-N
Condor
Cutlass
Diazinon
Furadan
Guthion
Insecticidal Soap

Lorsban
MPV
Omite
Orthene
Petroleum
   Oil-dormant
   and summer oils
Pyrellin
Pyrethrin
Sevin
Telone
Telone C-17

CRESS
Agree
Align
Condor
Cutlass
Permethrin

CUCUMBERS
Adios
Agree
Align
Asana
Augn
Bacillus thuringiensis
Biosafe
Condor
Cutlass
Diazinon
Furadan
Guthion
Insecticidal Soap
Kelthane
Lindane
Lorsban
Malathion
Methaldehyde
Methomyl
Methoxychlor
Methyl bromide
Mocap
Permethrin
Petroleum Oils

Pyrellin
Pyrethrin
Sevin
Spod-X
Telone
Telone C-17
Thiodan
Trigard
Vapam
Vendex
Vydate
Xentari

CURRANTS
Agree
Align
Bacillus thuringiensis
Condor
Cutlass
Malathion
Methoxychlor
Petroleum
   Oil-dormant
   and summer oils
Pyrellin
Pyrethrin
Telone
Telone C-17

DANDELION
Align
Bacillus thuringiensis
Condor
Cutlass
Insecticidal Soap
Methomyl
Permethrin
Sevin

DATES
Lorsban
Malathion
Methyl bromide
Mg Phosphide
Phostoxin

Pyrellin
Sulfur
Telone
Telone C-17

DILL
Azadirachtin
Condor
Cutlass
Pyrellin
Sevin
Xentari

EGGPLANT
Adios
Asana
Azadirachtin
Bacillus thuringiensis
Biosafe
Condor
Cutlass
Dibrom
Foil
Insecticidal Soap
Kryocide
Lindane
Malathion
Methaldehyde
Methomyl
Methoxychlor
Methyl bromide
MG Phosphide
Monitor
M-Trak
Nemacur
Permethrin
Petroleum Oils
Phostoxin
Pyrellin
Pyrethrin
Rotenone
Sevin
Telone
Telone C-17
Thiodan

Trident
Vendex
Vydate
Xentari

ENDIVE
Agree
Azadirachtin
Bacillus thuringiensis
Biosafe
Condor
Cutlass
Diazinon
Dimethoate
Insecticidal Soap
Larvin
Malathion
Methaldehyde
Methomyl
MG Phosphide
Permethrin
Pyrellin
Pyrethrin
Sevin
Telone
Telone C-17
Trigard
Xentari

ESCAROLE
Bacillus thuringiensis
Spod-X

FENNEL
Align
Pyrellin

FIGS
Azadirachtin
Lorsban
Malathion
Methyl bromide
Omite
Petroleum
    Oil-dormant
    and summer oils

Phostoxin
Pyrethrin
Sulfur
Telone C-17

FILBERTS
Asana
Bacillus thuringiensis
Condor
Cutlass
Guthion
Insecticidal Soap
Kelthane
Lorsban
Methyl bromide
Mg Phosphide
Permethrin
Petroleum Oils
Phostoxin
Pyrellin
Pyrethrin
Sevin
Telone C-17
Thiodan
Xentari

FLAX
Lindane
Malathion
Pyrethrins
Sevin
Telone
Telone C-17

GARLIC
Azadirachtin
Bacillus thuringiensis
Biosafe
Condor
Cutlass
Insecticidal Soap
Malathion
Methaldehyde
Methomyl
Methyl bromide
Nemacur

Permethrin
Pyrellin
Pyrethrin
Telone
Telone C-17
Vapam
Xentari

GENSING
Diazinon
Methaldehyde
Pyrellin

GINGER
Azadirachtin
Methyl Bromide
Pyrellin
Vydate

GOATS
Chlorpyrifos
Co-Ral
Ectrin
Methoxychlor
Permethrin
Pyrethrin

GOOSEBERRIES
Agree
Asana
Bacillus thuringiensis
Malathion
Methoxychlor
Pyrellin
Pyrethrin
Rotenone
Telone
Telone C-17

GRAPES
Agree
Align
Bacillus thuringiensis
Condor
Cutlass
Diazinon

Dibrom
Dimethoate
Enzone
Furadan
GBM Ropes
Guthion
Imidan
Insecticidal Soap
Kelthane
Kryocide
Lorsban
Malathion
Methaldehyde
Methomyl
Methoxychlor
Methyl bromide
MPV
Nemacur
Omite
Petroleum
  Oil-dormant
  and summer oils
Phostoxin
Pyrellin
Pyrethrin
Sevin
Telone
Telone C-17
Thiodan
Vapam
Vendex
Xentari

GRASS
Amdro
Asana
Astro
Baytex
Bacillus thuringiensis
Condor
Cutlass
Dibrom
Dimilin
L. giganteum
Logic

Malathion
Methaldehyde
Methomyl
Methoxychlor
Methyl parathion
Mycotrol
Orthene
Pyrellin
Sevin
Telone
Telone C-17
Thiodan

GUAR
Diazinon

GUAVA
Lorsban
Malathion
Pyrellin
Pyrethrins

HICKORY NUTS
Bacillus thuringiensis
Insecticidal Soap
Kelthane
Methyl bromide
Petroleum Oils
Pyrellin
Pyrethrin
Telone
Telone C-17

HOGS
Chlorpyrifos
Co-Ral
Dibrom
Ectrin
Methoxychlor
Permethrin
Pyrethrin
Rabon
Taktic

HOPS
Bacillus thuringiensis
Baythroid
Capture
Condor
Cutlass
Diazinon
Karate
Kelthane
Methomyl
Pyrellin
Omite
Telone
Telone C-17
Xentari

HORSES
Chlorpyrifos
DDVP
Dibrom
Ectrin
Methoxychlor
Permethrin
Pyrethrin
Rabon
Sevin
Tetralate

HORSERADISH
Bacillus thuringiensis
Biosafe
Condor
Cutlass
Malathion
Methaldehyde
Methomyl
MPV
Methyl bromide
Permethrin
Pyrellin
Pyrethrin
Sevin
Telone
Telone C-17

HUCKLEBERRIES
Agree
Asana
Bacillus thuringiensis
Pyrellin
Telone
Telone C-17

JOJOBA
Condor
Cutlass
Omite

KALE
Agree
Azadirachtin
Bacillus thuringiensis
Biosafe
Condor
Cutlass
Diazinon
Dibrom
Dimethoate
Insecticidal Soap
Kryocide
Lindane
Lorsban
Methaldehyde
Methomyl
Methoxychlor
MPV
Pyrellin
Pyrethrin
Sevin
Telone
Telone C-17
Thiodan
Xentari

KENAF
Telone C-17

KIWI
Actellic
Align

Bacillus thuringiensis
Condor
Cutlass
Diazinon
Imidan
Kryocide
Lorsban
Nemacur
Permethrin
Petroleum Oils
Pyrellin
Supracide
Xentari

KOHLRABI
Agree
Azadirachtin
Bacillus thuringiensis
Biosafe
Condor
Cutlass
Diazinon
Dibrom
Insecticidal Soap
Lorsban
Methaldehyde
Methoxychlor
MPV
Pyrellin
Pyrethrin
Sevin
Telone
Telone C-17
Thiodan
Xentari

KUMQUAT
Kelthane
Malathion
Methyl Bromide
MG Phosphide
Petroleum Oils
Telone

LEEKS
Bacillus thuringiensis
Biosafe
Condor
Cutlass
Insecticidal Soap
Malathion
Methaldehyde
Methomyl
Pyrellin
Pyrethrin
Telone
Telone C-17
Xentari

LENTILS
Asana
Azadirachtin
Bacillus thuringiensis
Condor
Cutlass
Dimethoate
Insecticidal Soap
Malathion
Methaldehyde
Methomyl
Methyl bromide
Pyrellin
Pyrethrin
Sevin
Xentari

LESPEDEZA
Bacillus thuringiensis
Malathion
Telone
Telone C-17

LETTUCE
Agree
Azadirachtin
Bacillus thuringiensis
Biosafe
Condor
Cutlass
Cypermethrin

Diazinon
Dibrom
Dimethoate
Di-Syston
Fury
Insecticidal Soap
Kryocide
Larvin
Lindane
Malathion
Methaldehyde
Methomyl
Methoxychlor
Methyl bromide
MG Phosphide
Monitor
MPV
Naturalis-L
Orthene
Permethrin
Petroleum Oils
Pyrellin
Pyrethrin
Sevin
Spod-X
Telone
Telone C-17
Thimet
Thiodan
Trigard
Vapam
Xentari

LOGAN
Pyrellin
Supracide

LUPINES
Dimethoate
Malathion

MACADAMIA NUTS
Azadirachtin
Bacillus thuringiensis
Insecticidal Soap

Malathion
Orthene
Pyrellin

MALANGA
Xentari

MANGOES
Align
Bacillus thuringiensis
Malathion
MG Phosphide
Petroleum Oils
Pyrellin
Pyrethrin
Supracide

MARIGOLDS
Align
Vendex

MELONS
  (Cantaloupes,
  Honeydew,
  Muskmelons, etc.)
Adios
Agree
Asana
Azadirachtin
Bacillus thuringiensis
Biosafe
Condor
Cutlass
Diazinon
Dibrom
Dimethoate
Furadan
Guthion
Insecticidal Soap
Kelthane
Kryocide
Lindane
Methaldehyde
Methomyl
Methoxychlor
Methyl bromide

Monitor
Naturalis-L
Permethrin
Petroleum Oils
Pyrellin
Pyrethrin
Rotenone
Sevin
Telone
Telone C-17
Thiodan
Trigard
Vydate
Xentari

MILLETS
Malathion
Methaldehyde
Methoprene
Mg Phosphide
Phostoxin
Pyrellin
Sevin

MINT
Azadirachtin
Bacillus thuringiensis
Dyfonate
Kelthane
Lorsban
Malathion
Methaldehyde
Methomyl
Omite
Orthene
Pyrellin
Pyrethrin
Telone C-17
Vapam
Vydate
Xentari

MUSHROOMS
Align
Apex

Armor
Diazinon
Dimilin
Lorsban
Malathion
MG Phosphide
Mocap
Pyrethrin

MUSTARD
Agree
Azadirachtin
Bacillus thuringiensis
Biosafe
Condor
Cutlass
Diazinon
Dibrom
Dimethoate
Insecticidal Soap
Malathion
Methaldehyde
Methomyl
MPV
Pyrellin
Pyrethrin
Sevin
Telone
Telone C-17
Thiodan
Vapam
Xentari

NECTARINES
Adios
Align
Apollo
Asana
Bacillus thuringiensis
Carzol
Codlure
Condor
Cutlass
Diazinon
Guthion

Imidan
Kelthane
Kryocide
Lorsban
Malathion
Meta-Systox-R
Methomyl
Methoxychlor
Methyl bromide
Omite
Petroleum
   Oil-dormant
   and summer oils
Pyrethrin
Sevin
Supracide
Telone
Telone C-17
Thiodan
Xentari

OATS
Bacillus thuringiensis
Condor
Cutlass
Diacon
Furadan
Guthion
Lindane
Lorsban
Malathion
Methaldehyde
Methomyl
Methoprene
Methoxychlor
Methyl bromide
Methyl parathion
Mg Phosphide
Mycotrol
Phostoxin
Pyrethrin
Reldan
Thiodan
Xentari

OKRA
Asana
Bacillus thuringiensis
Condor
Cutlass
Insecticidal Soap
Kryocide
Lindane
Malathion
Methaldehyde
Methyl bromide
Mocap
Nemacur
Pyrellin
Pyrethin
Sevin
Telone
Telone C-17

OLIVES
Align
Bacillus thuringiensis
Methaldehyde
Petroleum Oils
Pyrethrin
Sevin
Supracide
Telone
Telone C-17

ONIONS
Ammo
Azadirachtin
Bacillus thuringiensis
Biosafe
Condor
Cutlass
Diazinon
Dyfonate
Guthion
Insecticidal Soap
Lindane
Lorsban
Malathion
Methaldehyde

Methomyl
Methyl bromide
Permethrin
Pyrellin
Pyrethrin
Spod-X
Telone
Telone C-17
Vapam
Vydate
Xentari

ORNAMENTALS
Agree
Altosid
Amdro
Asana
Astro
Avid
Award
Azadirachtin
Azatin
Bacillus thuringiensis
Basamid
Biosafe
Chloropicrin
Citation
Condor
Cutlass
Diazinon
DDVP
Dibrom
Dimethoate
Dimilin
Di-Syston
Dylox
Dursban
Enstar
Exhibit
Fenvalerate
Ficam
Fireban
Furadan
Guthion
Imidan

Insecticidal soap
Kelthane
Kryocide
Logic
Malathion
Margosan-O
Mavrik
Merit
Methaldehyde
Methoxychlor
Methyl bromide
Minex
Mocap
M-One
Morestan
Naturalis-L
Nemacur
Nosema spore
Oftanol
Omite
Ornamite
Orthene
Oxamyl
Pageant
Pentac
Permethrin
Petroleum
  Oil-dormant
  and summer oils
Plant Fume
Precision
Pyrellin
Pyrethrin
Rotenone
Scimitar
Sevin
Spod-X
Sumithrin
Supracide
Talstar
Tame
Telone
Telone C-17
Temik
Tempo

Tetralate
Thiodan
Trident
Turcam
Vapam
Xentari

PAPAYAS
Align
Bacillus thuringiensis
Malathion
Methyl Bromide
MG Phosphide
Permethrin
Petroleum Oils
Pyrellin
Vendex

PARSLEY
Azadirachtin
Bacillus thuringiensis
Biosafe
Condor
Cutlass
Diazinon
Insecticidal Soap
Methomyl
Permethrin
Phosdrin
Pyrellin
Sevin
Xentari

PARSNIPS
Azadirachtin
Bacillus thuringiensis
Biosafe
Condor
Cutlass
Diazinon
Insecticidal Soap
Malathion
Methaldehyde
Methyl bromide
Nicotine

Pyrellin
Pyrethrin
Sevin
Telone
Telone C-17

PASSION FRUIT
Malathion

PEACHES
Adios
Agree
Apollo
Asana
Azadirachtin
Bacillus thuringiensis
Carzol
Codlure
Condor
Cutlass
Diazinon
Dibrom
Guthion
Imidan
Insecticidal Soap
Isomate-M
Kelthane
Lorsban
Malathion
Methaldehyde
Methomyl
Methoxychlor
Methyl bromide
Nemacur
Omite
Permethrin
Petroleum
   Oil-dormant
   and summer oils
Pyrellin
Pyrethrin
Sevin
Supracide
Telone
Telone C-17

Thiodan
Vapam
Vendex
Vydate
Xentari

PEANUTS
Actellic
Adios
Asana
Azadirachtin
Bacillus thuringiensis
Comite
Condor
Cutlass
Diacon
Di-Syston
Dyfonate
Furadan
Kryocide
Lorsban
Methaldehyde
Methomyl
Methoxychlor
Methyl bromide
Mg Phosphide
Mocap
MPV
Naturalis-L
Nemacur
Omite
Orthene
Phostoxin
Pyrellin
Pyrethrin
Sevin
Spod-X
Telone
Telone C-17
Temik
Thimet
Vydate
Xentari

PEARS
Amitraz
Apollo
Asana
Azadirachtin
Bacillus thuringiensis
Carzol
Codlure
Condor
Cutlass
Diazinon
Dimethoate
Guthion
Imidan
Insecticidal soap
Kelthane
Lorsban
Malathion
Methaldehyde
Methomyl
Methoxychlor
Methyl bromide
Morestan
Omite
Permethrin
Petroleum
   Oil-dormant
   and summer oils
Pyrellin
Pyrethrin
Rotenone
Ryanodine
Savey
Sevin
Supracide
Telone
Telone C-17
Thiodan
Vapam
Vendex
Vydate
Xentari

24

PEAS
Adios
Agree
Asana
Azadirachtin
Bacillus thuringiensis
Condor
Cutlass
Dibrom
Dimethoate
Di-Syston
Imidan
Insecticidal Soap
Kryocide
Lorsban
Methaldehyde
Methomyl
Methoxychlor
Methyl bromide
Pyrellin
Pyrethrin
Sevin
Spod-X
Telone
Telone C-17
Xentari

PECANS
Align
Asana
Bacillus thuringiensis
Condor
Cutlass
Cypermethrin
Dimethoate
Di-Syston
Fumi-cel
Fury
Guthion
Imidan
Insecticidal Soap
Lorsban
Malathion
Methomyl

Methyl bromide
MG Phosphide
Petroleum Oils
Phostoxin
Pyrellin
Pyrethrin
Sevin
Supracide
Telone
Telone C-17
Temik
Thiodan
Vendex
Xentari

PEPPERS
Adios
Agree
Asana
Azadirachtin
Bacillus thuringiensis
Biosafe
Condor
CutlasAs
Diazinon
Dibrom
Dimethoate
Disyston
Dyfonate
Furadan
Insecticidal soap
Kelthane
Kryocide
Lindane
Lorsban
Malathion
Methaldehyde
Methomyl
Methoxychl
Methyl bromide
MG Phosphide
Monitor
MPV
Nemacur
Orthene

Permethrin
Petroleum Oils
Pyrellin
Pyrethrin
Sevin
Spod-X
Telone
Telone C-17
Thiodan
Trigard
Vydate
Xentari

PERSIMMON
Align
Condor
Cutlass
MG Phosphide
Pyrellin
Telone
Telone C-17

PINEAPPLE
Align
Bacillus thuringiensis
Condor
Cutlass
Diazinon
Methyl bromide
Mocap
Nemacur
Pyrellin
Sevin
Telone
Telone C-17
Thiodan
Vydate
Xentari

PISTACHIO
Azadirachtin
Bacillus thuringiensis
Guthion
Imidan
Insecticidal Soap

Mg Phosphide
Permethrin
Petroleum Oils
Phostoxin
Pyrellin
Sevin

PLAINTAINS
Align
MG Phosphide
Petroleum Oils

PLUMS
Asana
Azadirachtin
Bacillus thuringiensis
Carzol
Codlure
Condor
Cutlass
Diazinon
Guthion
Imidan
Kelthane
Kryocide
Lorsban
Methoxychlor
Methyl bromide
Morestan
Omite
Petroleum
   Oil-dormant
   and summer oils
Pyrellin
Pyrethrin
Sevin
Supracide
Telone
Thiodan
Vapam
Vendex
Xentari

POMEGRANATES
Condor
Cutlass

Guthion
Methomyl
Pyrellin
Telone
Telone C-17

POTATOES
Adios
Admire
Agree
Asana
Azadirachtin
Bacillus thuringiensis
Biosafe
Comite
Condor
Cutlass
Diazinon
Dimethoate
Di-Syston
Dyfonate
Foil
Furadan
Guthion
Imidan
Insecticidal Soap
Malathion
Methaldehyde
Methomyl
Methoxychlor
Methyl bromide
Mocap
Monitor
MPV
M-Trak
Mycotrol
Novador
Omite
Permethrin
Petroleum Oils
Phosphamidon
Pyrellin
Pyrethrin
Rotenone
Sevin

Spod-X
Telone
Telone C-17
Tennax
Thimet
Thiodan
Trident
Vapam
Vydate
Xentari

POULTRY
Altosid
Boric Acid
Chlorpyrifos
DDVP
Diazinon
Dibrom
Larvadex
Methoxychlor
Permethrin
Pyrethrin
Rabon
Sevin

PRUNES
Align
Asana
Bacillus thuringiensis
Carzol-SP
Codlure
Condor
Cutlass
Diazinon
Guthion
Imidan
Kryocide
Lorsban
Methoxychlor
Methyl bromide
Morestan
Omite
Petroleum
   Oil-dormant
   and summer oils

Pyrethrin
Sevin
Supracide
Telone
Telone C-17
Thiodan
Vapam
Vendex
Xentari

PUMPKINS
Adios
Agree
Asana
Azadirachtin
Bacillus thuringiensis
Biosafe
Condor
Cutlass
Dibrom
Furadan
Insecticidal Soap
Kelthane
Kryocide
Lindane
Lorsban
Methaldehyde
Methomyl
Methoxychlor
Methyl bromide
Permethrin
Petroleum Oils
Pyrellin
Pyrethrin
Rotenone
Sevin
Telone
Telone C-17
Thiodan
Trigard
Vapam
Vydate

QUINCE
Bacillus thuringiensis
Codlure

Condor
Cutlass
Guthion
Kelthane
Kryocide
Malathion
Methyl bromide
Methoxychlor
Petroleum
  Oil-dormant
  and summer oils
Pyrellin
Sevin
Telone
Telone C-17

RADICCHIO
Agree
Diazinon
Pyrellin

RADISHES
Asana
Azadirachtin
Bacillus thuringiensis
Biosafe
Condor
Cutlass
Diazinon
Dyfonate
Insecticidal Soap
Lorsban
Malathion
Methaldehyde
Methoxychlor
Methyl bromide
MPV
Petroleum Oils
Pyrellin
Pyrethrin
Sevin
Telone
Telone C-17
Vapam

RAPE (canola)
Condor
Cutlass
Mycotrol
Thiodan

RHUBARB
Permethrin
Pyrellin

RICE
Actellic
Altosid
Bacillus thuringiensis
BTI
Condor
Copper sulfate
Cutlass
Furadan
L. giganteum
Malathion
Methoprene
Methoxychlor
Methyl bromide
Methyl parathion
Mg Phosphide
Phostoxin
Pyrellin
Pyrethrin
Pyrenone
Reldan
Sevin

RUTABAGAS
Agree
Azadirachtin
Bacillus thuringiensis
Biosafe
Condor
Cutlass
Diazinon
Insecticidal Soap
Kryocide
Lorsban
Methaldehyde

Methoxychlor
Methyl bromide
MPV
Pyrellin
Pyrethrin
Sevin
Telone
Telone C-17

RYE
Bacillus thuringiensis
Condor
Cutlass
Guthion
Lindane
Malathion
Methomyl
Methoprene
Methoxychlor
Methyl bromide
Methyl parathion
Mg Phosphide
Phostoxin
Pyrethrin
Sevin
Telone
Telone C-17
Thiodan
Xentari

SAFFLOWER
Condor
Cutlass
Dibrom
Dimethoate
Lindane
Mycotrol
Phostoxin
Supracide
Telone
Telone C-17
Thiodan
Xentari

SAINFOIN
Malathion

SALSIFY
Bacillus thuringiensis
Biosafe
Condor
Cutlass
Malathion
Methaldehyde
Methyl bromide
MG Phosphide
Pyrellin
Pyrethrin
Sevin
Telone
Telone C-17

SAPOTE
Lorsban
Pyrellin

SESAME
Phostoxin

SHALLOTS
Bacillus thuringiensis
Biosafe
Condor
Cutlass
Insecticidal Soap
Malathion
Methaldehyde
Pyrellin
Pyrethrin
Telone
Telone C-17

SHEEP
Chlorpyrifos
Co-Ral
Diazinon
Dibrom
Ectrin
Methoxychlor

Permethrin
Pyrethrin

SORGHUM (MILO)
Actellic
Align
Bacillus thuringiensis
Comite
Condor
Counter
Cutlass
Diacon
Dimethoate
Di-Syston
Furadan
Gaucho
Lindane
Lorsban
Methomyl
Methoprene
Methoxychlor
Methyl bromide
Methyl parathion
Mg Phosphide
Parathion
Phostoxin
Pyrellin
Pyrethrin
Reldan
Sevin
Supracide
Temik
Thimet

SOYBEANS
Asana
Bacillus thuringiensis
Condor
Cutlass
Design
Diazinon
Dimethoate
Dimilin
Di-Syston
Furadan

Larvin
Lindane
L. giganteum
Lorsban
Methomyl
Methoxychlor
Methyl parathion
Mg Phosphide
Methaldehyde
MPV
Mycotrol
Parathion
Permethrin
Phostoxin
Pyrellin
Scout
Sevin
Slam
Spod-X
Telone
Telone C-17
Temik
Thimet
Vydate
Xentari

SPICES
Azadirachtin
Cutlass
Ethylene Oxide
Insecticidal Soap
Pyrellin

SPINACH
Agree
Azadirachtin
Bacillus thuringiensis
Biosafe
Condor
Cutlass
Diazinon
Dimethoate
Enstar
Insecticidal Soap
Larvin

Lindane
Malathion
Methaldehyde
Methomyl
Methoxychlor
Permethrin
Pyrethrin
Sevin
Telone
Telone C-17
Thiodan
Trigard
Vapam
Xentari

SQUASH
Adios
Agree
Asana
Azadirachtin
Bacillus thuringiensis
Biosafe
Condor
Cutlass
Diazinon
Dibrom
Furadan
Insecticidal Soap
Kelthane
Kryocide
Lindane
Malathion
Methaldehyde
Methomyl
Methoxychlor
Methyl bromide
Permethrin
Petroleum Oils
Pyrellin
Pyrethrin
Rotenone
Sevin
Telone
Telone C-17
Thiodan

Trigard
Vapam
Vydate
Xentari

STRAWBERRIES
Agree
Agri-mek
Align
Bacillus thuringiensis
Biosafe
Condor
Cutlass
Diazinon
Dibrom
Di-Syston
Dyfonate
Furadan
Guthion
Insecticidal Soap
Kelthane
Lorsban
Malathion
Methaldehyde
Methomyl
Methoxychlor
Methyl bromide
Morestan
Nemacur
Omite
Petroleum
   Oil-dormant
   and summer oils
Pyrethrin
Sevin
Telone
Telone C-17
Thiodan
Vapam
Vendex
Xentari

SUGAR APPLE
Supracide

SUGAR BEETS
Align
Bacillus thuringiensis
Counter
Condor
Cutlass
Diazinon
Dibrom
Dyfonate
Enstar
Furadan
Insecticidal Soap
Lorsban
Malathion
Methomyl
Methyl bromide
Methyl parathion
Monitor
Mycotrol
Petroleum Oils
Pyrellin
Sevin
Spod-X
Telone
Telone C-17
Temik
Thimet
Thiodan
Xentari

SUGARCANE
Asana
Diazinon
Furadan
Guthion
Lorsban
Mocap
Telone
Telone C-17
Temik
Thimet
Thiodan

SUNFLOWER
Asana
Bacillus thuringiensis
Condor
Cutlass
Furadan
Lindane
Lorsban
Methyl parathion
Mg Phosphide
Mycotrol
Parathion
Permethrin
Phostoxin
Pyrellin
Sevin
Spod-X
Supracide
Thiodan
Xentari

SWEET POTATOES
Adios
Azadirachtin
Bacillus thuringiensis
Biosafe
Condor
Cutlass
Diazinon
Insecticidal Soap
Lorsban
Magnesium phosphide
Malathion
Methaldehyde
Methoxychlor
Methyl bromide
Mocap
Pyrellin
Pyrethrin
Sevin
Spod-X
Telone
Telone C-17
Thiodan
Vydate

SWISS CHARD
Agree
Align
Bacillus thuringiensis
Biosafe
Condor
Cutlass
Diazinon
Dimethoate
Insecticidal Soap
Methaldehyde
Methomyl
Permethrin
Pyrellin
Pyrethrin
Sevin
Telone
Telone C-17
Trigard

TOBACCO
Agree
Align
Bacillus thuringiensis
Basamid
Condor
Cutlass
Dianex
Dimethoate
Di-Syston
Dyfonate
Furadan
Guthion
Insecticidal Soap
Methomyl
Mocap
M-Peril
MPV
Nemacur
Orthene
Petroleum Oils
Sevin
Spod-X
Spur
Supracide

Telone C-17
Thiodan
Vydate
Xentari

TOMATOES
Adios
Agree
Agri-mek
Asana
Azadirachtin
Bacillus thuringiensis
Biosafe
Checkmate TPW
Condor
Cutlass
Diazinon
Dimethoate
Di-Syston
Dyfonate
Enstar
Foil
Guthion
Insecticidal soap
Kelthane
Kryocide
Lindane
Lorsban
Lycolure
M-Trak
Malathion
Methaldehyde
Methomyl
Methoxychlor
Methyl bromide
MG Phosphide
Monitor
MPV
Naturalis-L
Permethrin
Petroleum Oils
Pyrellin
Pyrethrin
Rotenone
Sevin

Telone
Telone C-17
Thiodan
Trident
Vapam
Vydate
Xentari

TREFOIL
Bacillus thuringiensis
Guthion
Malathion
Methaldehyde
Methyl parathion
Pyrethrin
Sevin
Telone
Telone C-17

TURF
Affirm
Amdro
Astro
Award
Azadirachtin
Azatin
Basamid
Baygon
Biosafe
Crusade
Diazinon
Doom
Dursban
Dyfonate
Dylox
Exhibit
Ficam
Furadan
Kelthane
Lindane
Logic
Mainstay
Malathion
Mavrik
Merit

Methaldehyde
Methomyl
Mocap
Morestan
Naturalis-L
Nemacur
Oftanol
Orthene
Pageant
Proxol
Pyrethrin
Scimitar
Sevin
Talstar
Telone
Telone C-17
Tempo
Triumph
Turcam
Vapam
Yardex
Xentari

TURNIPS
Agree
Asana
Azadirachtin
Bacillus thuringiensis
Condor
Cutlass
Diazinon
Dyfonate
Insecticidal Soap
Lorsban
Malathion
Methaldehyde
Methomyl
Methoxychlor
Methyl bromide
MPV
Permethrin
Pyrellin
Pyrethrin
Sevin
Telone

Telone C-17
Thiodan
Vapam

VETCH
Bacillus thuringiensis
Malathion
Methaldehyde
Methyl parathion
Pyrethrin
Telone
Telone C-17

WALNUTS
Apollo
Asana
Azadirachtin
Bacillus thuringiensis
Codlure
Condor
Cutlass
Diazinon
Dibrom
Dimilan
Ethylene Oxide
Guthion
Insecticidal Soap
Kelthane
Lorsban
Malathion
Methoprene
Methyl bromide
Mg Phosphide
Morestan
Omite
Permethrin
Petroleum Oils
Phostoxin
Pyrellin
Pyrethrin
Supracide
Sevin
Telone
Telone C-17

Thiodan
Vendex

WATERCRESS
Align
Bacillus thuringiensis
Condor
Cutlass
Malathion
MPV

WATERMELON
Adios
Agree
Asana
Azadirachtin
Bacillus thuringiensis
Biosafe
Condor
Cutlass
Diazinon
Dimethoate
Guthion
Insecticidal Soap
Kelthane
Kryocide
Lindane
Methaldehyde
Methomyl
Permethrin
Petroleum Oils
Pyrellin
Pyrethrin
Sevin
Telone
Telone C-17
Thiodan
Trigard
Vorlex
Vydate
Xentari

WHEAT
Actellic
Bacillus thuringiensis

Condor
Cutlass
Diacon
Dimethoate
Di-Syston
Furadan
Guthion
Lindane
Malathion
Methaldehyde
Methomyl
Methoxychlor
Methyl bromide
Methyl parathion
Mg Phosphide
Mycotrol
Parathion
Phostoxin
Pyrellin
Pyrethrin
Reldan
Sevin
Spod-
Thimet
Thiodan

WILD RICE
Malathion

INSECTICIDES

ACEPHATE, ORTHENE,　　　　　　　Mfg: Valent
PAYLOAD, OTTO

　Ants, aphids, armyworms, aster leaf hopper, bagworm, bean leafbeetle, birch leafminers, blackgrass bug, blossom worms, bollworms, boxelder bug, budworms, brown cranberry spanworm, cockroaches, European corn borer, blackfly, cabbage looper, California oak moth, cankerworms, chinch bug, cockroaches, confused flour beetle, corn earworm, cranberry, crickets, Cuban laural thrips, cutworms, Douglas fir tussock moth, earwigs, Eastern tent caterpillar, European corn borer, Elmleaf beetle, fall webworm, fir brats, leabeetle, fleahoppers, forest tent caterpillar, grasshoppers, greenbug, green loverworm, gypsy moth, hornworms, Indian meal moth, Japanese beetle, lace bugs, leafhoppers, leafminers, leafrollers, loopers, lygus, mealybugs, mirids, Mormon crickets, Nantucket pine tip moth, omnivorous leafrollers, pillbug, pink bollworm, plant bugs, Ponderosa pineneedle miner, red-humped caterpillar, root weevils, sawflies, scales, silver spotted skiper, stinkbug, sowbug, sod webworm, three-cornered alfalfa hopper, thrips, trogoderma, tobacco budworm, velvet bean caterpillar, wasps, webworms, and whiteflies.

ALDICARB, TEMIK　　　　　　　Mfg: Rhone Poulenc

　Aphids, boll weevils, citrus mites, Colorado potato beetle, cotton leaf perforator, fleahoppers, leafhoppers, leafminers, lygus, mealybugs, Mexican bean beetle, mites, nematodes, plant bugs, root maggots, three cornered alfalfa hopper, thrips, whiteflies.

ALUMINUM PHOSPHIDE,　　　　　Mfg: Degesch, Pestcon
PHOSTOXIN, FUMITOXIN　　　　　System and others

　Stored grain insects.

AMITRAZ, MITAC, TAKTIC,　　　　Mfg: Agr Evo
OVASYN

　Mites, pear psylla, bollworm, tobacco budworm, pink bollworm, whiteflies.

APM Ropes (pheromone)　　　　Mfg: Mitsubishi

　Artichoke plume moth.

AVERMECTIN B, AFFIRM, AGRI-MEK, AVID, ZEPHYR

Mfg: MSD Ag Vet

Fireants, mites, and leafminers.

ALIGN, AZADIRACHTIN, AZATIN, BENEFIT, MARGOSAN-O, TURPLEX, NEEMISIS, ALIGN, NEEMIX, BIONEEM

Mfg: W.R. Grace, Agri-Dyne, Ringer

Whiteflies, mealybugs, thrips, gypsy moth, aphids, loopers, caterpillars, leafminers, fungus gnats, armyworms, cutworms, leafrollers, leafhoppers, webworms, spruce budworms, sawflies, soft brown scale, weevils, hornworms, psyllids, leafminers, whiteflies, thrips, mealybugs, aphids, fruitflies, armyworms, Mexican bean beetle, weevils, webworms, elm leaf beetle, bagworms, tent caterpillers, grasshoppers, craneflies, mole crickets, white grubs, chinch bugs, hyperodes weevils, billbugs, lacebugs, boxelder bugs, sawfly, fungus gnats, diamondback moth, cabbageworms, melonworms, pickleworms, corn earworms, tomato fruitworms, bollworms, borers, squash bugs, pinworms, Colorado potato beetle, European corn borer, cucumber beetle, vine borers, cutworms, leafrollers, grapeleaf skeletonizer, twig girdlers, codling moth, pepper weevil, onion maggots.

AZINPHOS-METHYL, GUTHION

Mfg: Miles Inc. and Gowan

Alfalfa plant bug, alfalfa weevil, American plum borer, aphids, apple maggot, black vine weevil, blueberry maggot, boll weevil, bollworm, brown cotton leafworms, cabbage maggot, carrot weevil, cherry fruitfly, cherry leafminer, citrus rootworm, citrus thrips, clover leafweevil, codling moth, Colorado potato beetle, cone midge, cone moth, cone worm, cotton leafhopper, cotton leafworm, current borer, diamondback moth, European apple sawfly, European brown snail, European fruit lecarnium, European pine shootmoth, European red mite, eye-spotted bud moth, false webworm, flea beetle, fleahoppers, filbert worm, Forbes scale, fruit flies, fruit tree leaf roller, fruitworms, grape berry moth, grape bud beetle, grape cane girdlers, hickory shuckworm, imported cabbageworm, hornworm, lacebugs, leafhoppers, leaf miners, leaf rollers, lesser clover leaf weevil, lesser peach twig borer, lygus, May beetles, mealy bug, Mexican bean beetle, Mineola moth, mites, Nantucket pine tip moth, navel orangeworm, oblique-banded leafroller, orange tortrix, Oriental fruitmoth, peach treeborer, peach twigborer, pecan casebearer, pink bollworm, plum curculio, putnam scale, rapid plant bug, raspberry rootborer, redbanded leafroller, scales, seedworms, Southern green stinkbug, spittlebug, stinkbug, strawberry leaf roller, tarnished plant bug, thrips, tipworm, tobacco budworm, tobacco hornworm, tobacco flea beetle, tubermoth, Tussock moth, twig girdlers, webworms.

BACILLUS-THURINGIENSIS, VAR
KURSTAKI, VAULT, LEPID, RAPTOR
BIOBIT, DIPEL, FORAY, MVP,
JAVELIN,THURICIDE

Mfg: Abbott, Sandoz,
Novo, DuPont, Mycogen,
Ecogen, American Cyan-
amid and others

Alfalfa caterpillar, almond moth, Amorbia moth, armyworms, artichoke plume moth, bagworm, banana skipper, bollworm, budworms, Cabbage looper, California oak moth, cankermoths, cankerworm, celery leaftier, celery looper, citrus cutworm, codling moth, cotton leafworm, cutworms, diamond-back moth, difoliatry caterpillar, Douglas fir tussock moth, elm spanworm, European cornborer, European skipper, fall webworm, filbert leafroller, forest tent caterpillar, fruit tree leafroller, grape leaf folder, Great Basin tent caterpillar, green cloverworm, gypsy moth, hornworms, imported cabbageworm, Indian meal moth, loopers, navel orange worm, oakmoth, olecinder moth, omnivorous leaf roller, omnivo-rous looper, orange dog, oriental fruit moth, peach twig borer, pine butterfly, podworms, range caterpillar, red-banded leafroller, red-humped caterpillar, rindworm, spanworms, tomato pinworm, tent caterpillar, salt marsh caterpillar, sorghum headworm, soybean looper, spruce budworm, sunflower moth, tent caterpillar, tobacco budworm, tobacco hornworm, tobacco moth, tomato fruitworm, tufted apple budmoth, tussolk moth, varigated leaf roller, velvet bean caterpillar, wax moth larvae.

BEAUVARIA BASSIANA, NATURALIS-L,
MYCOTROL-GH-OH

Mfg: Troy Chemical Co.,
Mycotech Corp.

Boll weevil, whiteflies, fleahoppers, aphids, thrips, mites, leaf feeding caterpillers, lygus bugsm cucumber beetles, elm leaf beetle, leafhoppers, strawberry root weevil, obscure root weevil, blackvine root weevil, chinch bugs, crickets, grasshoppers, psyllids, mormon crickets, locusts

BENDIOCARB, FICAM, ROTATE,
TURCAM, DYCARB

Mfg: Agr Evo

Ants, aphids, Azalea caterpillar, Azalea leaf miner, bagworms, bees, billbugs, black vine weevil, bronze birch borer, carpet beetles, centipedes, chinch bugs, citrus blackfly, clothes moth, cockroaches, crickets, drugstore beetle, earwigs, elm leaf beetle, European chafer, fireants, firebrats, fleas, flies, flour beetles, grain beetles, green June bugs, ground beetles, gypsy moth, Japanese beetle, lacebugs, leafhoppers, lesser grain borer, mealybugs, millipedes, molecrickets, mosquitoes, Nantucket pinetip moth, Northern masked chafer, northern pine weevil, obscure rootweevil, oleander caterpillar, Pales pine weevil, pantry pest, pillbugs, pine spittlebug, pine tipmoth, poplar tentmakers, rice weevils, scales, scorpions, silverfish, sowbugs, spider beetles, spiders, spring cankerworm, tent caterpillars, termites, thrips, ticks, tobacco moth, wasps, webworms, white grub, willow leafbeetle, whiteflies, yellow-necked caterpillar.

BIFENTHRIN, BRIGADE, CAPTURE,   Mfg: FMC
TALSTAR, BIFLEX

Aphids, armyworms, boll weevil, bollworms, cabbage looper, cotton fleahopper, cotton leaf perforator, cutworms, European corn borer, leafhoppers, lygus. mealybugs, omnivorous leafrollers, pink bollworm, plant bugs, saltmarsh centerpillar, scales, stink bugs, thrips, tobacco budworm, two-spotted spider mites, white flies, bagworms, black vine weevil, brood mites, chinch bugs, corn earworm, cucumber beetle, elm leaf beetle, webworms, fleabettles, planthoppers, fungus gnat grasshoppers, Japanese beetle, lacebugs, leafminers, leafrollers, mirid bugs, oriental fruitmoth, pine tip moth twig borers.

BIO SAFE, VECTOR (beneficial nematodes),   Mfg: Biosys

Cutworms, white grubs, cucumber beetle, onion maggots, sod webworms, blackvine weevil, June beetle, Oriental beetle, rose chafer, strawberry root weevil, billbugs, cabbage root maggot, root maggots, wireworm, Japanese beetle, flea beetle, armyworm, carrot weevil, mole crickets, Asiatic garden beetle, leather jackets, European chafer, masked chafer, citrus weevils, whitefringed beetles, cranberry girdler, and leatherjacket larvae.

BIOVECTOR (beneficial nematodes)   Mfg: Biosys

Sugarcane borer, blue green weevil.

BORIC ACID   Mfg: Numerous

Ants, cockroaches, darkling ground beetle, hide beetle, lesser mealworm beetle, palmetto bugs, silverfish, waterbugs.

B.t. AIZAWAI/KURSTATI,   Mfg: Ciba
DESIGN

Bollworms, armyworms, tobacco budworm, loopers, podworms, velvetbean caterpiller.

B.t. strain EG 2371, CUTLASS   Mfg: Ecogen

Armyworms, cabbage looper, cabbage webworm, cross striped cabbageworm, diamondback moth, imported cabbageworm, tobacco budworm, European corn borer, Southwestern corn borer, bollworm, cotton leaf perforator, saltmarsh caterpillar, leafrollers, omnivorous leaftier cutworms, green clover worm, podworms, velvetbean caterpillar, banded sunflower moth, headmoth, sunflower moth, tobacco hornworm, azalea moth, Ello moth, oleander moth bagworm, blackheaded budworm, browntail moth, California oakworm, Douglas fir tussock moth, elm spanworm, fall web worm, fruit tree leafroller, green striped maple worm, Eastern tent caterpillar, leafrollers, Navel orangeworm, oriental fruit moth, peach twig borer, tufted apple budmoth, walnut caterpiller, filbert leafroller, amboria, orange dog, sphinx moth, grape berry moth, cherry fruitworm, grape leaf folder, grapeleaf skeletonizer, green fruitworm, spanworm, banana skipper, gypsy moth, pine budworm, minosa webworm, pine butlerfly, red humped caterpillar, cankerworms, saddle prominent caterpillar, spruce budworm, tent caterpillar, fortrex Western tassock moth, artichoke plume moth, webworms, celery leaftier, corn ear worm, melonworm, pickleworm, tomato hornworm, tomato pinworm, European skipper.

B.t. strain EG 2424, FOIL   Mfg: Ecogen

Colorado potatoe beetle, European corn borer, armyworm, loopers.

B.t. strain 2348, CONDOR   Mfg: Ecogen

Gypsy moth, spruce budworm, soybean looper, velvet bean caterpiller, green clover worm, European corn borer, southwestern corn borer, tobacco budworm.

B.t. Var AIZAWAI, XENTARI   Mfg: Abbott

Alfalfa caterpillar, armyworms, European skipper, loopers, cutworms, hornworms, grape berry moth, grape leafroller, grape leaf skeletonizer leafrollers, loopers, melonworms, orange tortrex, saltmarsh caterpillar, tobacco budworm, cabbageworm, diamondback moth, green cloverworm, Heliothis, webworms, melonworms, rindworms, Azalea caterpiller, Ello moth, Io moth, oleander moth, fruitworms, podworms, velvetbean caterpiller, cankerworms, red humped caterpiller, tent caterpillers, walnut caterpillers, coddling moth, gypsy moth, oriental fruit moth, tufted apple budmoth, twig borer, spanworm, orangedog, head moth.

B.t. Var AIZAWAI strain GC-91, **AGREE**   **Mfg: Ciba**

Loopers, imported cabbageworm, diamondback moth, armyworms, tomato fruit worm, hornworm, tobacco budworm, peachtwig borer, grapeleaf skeletonizer, European corn borer, rindworm complex, melonworm.

B.t. Var ISRALENSIS, **BACTIMOS,**   **Mfg: Abbott, Sandoz,**
TEKNAR,VECTOBAC, SKEETAL,   **Novo, American**
ACROBE   **Cyanamid**

Blackflies, mosquitoes.

B.t. Var KURSTAKI ENCAPSULATED,   **Mfg: Mycogen**
M-PERIL, MVP

European corn borer, diamondback moth, loopers, Southwestern corn borer, imported cabbageworm, armyworms, fall armyworm, hornworm, cutworm, leafrollers, tobacco budworms, western grapeleaf skeletonizer, grapeberry moth, velvetbean caterpiller, green cloverworm.

B.t. Var TENEBRIONIS, **TRIDENT II,**   **Mfg: Sandoz/Novo/**
NOVODOR, M-TRAK   **Mycogen**

Colorado potato beetle, elm leaf beetles.

CARBARYL, **SEVIN**   **Mfg: Rhone Poulenc**

Alfalfa caterpillar, alfalfa looper, alfalfa weevil, ants, Apache cicada, aphids, apple maggot, apple mealybug, apple rust mite, apple sucker, armyworms, asparagus beetles, aster leafhopper, bagworm, bean leafbeetle, bean leafroller, bedbugs, bees, birch leaf miner, black cherry aphid, black scale, blister beetles, blueberry maggot, bluegrass billbug, body lice, boll weevil, bollworm, boxelder bug, boxwood leafminer, brown dog tick, brown soft scales, cabbage looper, calico scale, California red scale, cankerworms, catfacing insect, cherry fruit fly, cherry fruitworm, chicken mite, chiggers, chinch bugs, citricola scale, citrus cutworm, climbing cutworm, clover head weevil, cockroaches, codling moth, Colorado potato beetle, corn earworm, corn rootworm, cotton fleahopper, cotton leaf perforator, cowpea curculio, crab louse, cranberry fruitworm, crickets, cucumber beetle, cutworms, darkling ground beetles, deer ticks, diamondback moth, earwigs, Eastern apple sawfly, Eastern spruce gall aphid, Eastern tent caterpillar, elm leaf beetle, eriophyid mites, European apple sawfly, European chafer, European corn borer, European earwig, European fruit lecanium, eye-spotted bud moth, fall armyworm, fall cankerworm, filbert leaf roller, filbert moth, filbert worm, fireworms, flea beetle, fleas, Forbes scale, forest tent caterpillar, fowl tick, frosted scale, fruit tree leaf roller, fruitworms, grape berry moth, grape leaffolder, grape leafhopper, grape leaf skeletonizer, grasshoppers, great basin tent caterpillar, green cloverworm, green fruitworms, green June beetle, green June bug, gypsy moth, Harlequin bug, head lice, hornets, hornworm, housefly, imported cabbageworm, imported fire ant, Japanese beetle, June beetles, lace bugs, leafhoppers, leaf rollers, lecanium scales, Lepidopterous larvae, lesser peach tree borer, lice, lima bean potborers, lygus bugs, maple leafcutter, meadow spittlebug, mealybugs, mealy plum aphid, melonworm, millipedes, mimosa webworm, Mexican bean beetle, mose crickets, mosquitoes, moth flies, navel orangeworm, Northern fowl mite, oak leaf miners, olive scale, omnivorous leaftier, omnivorous leafroller, orange tortrix, Oriental fruit moth, oyster shell scale, pandemis moth, peach twigborer, pear leaf blister moth, pear psylla, pear rust mite, pickleworm, pink bollworm, pinworms, plant bugs, plum curculio, prune leafhopper, psyllids, raspberry aphids, redbanded leafroller, rosebug, salt marsh caterpillar, San Jose scale, sand flies, sap beetles, sawflies, scales, six-spotted leafhopper, snowy tree crickets, sod webworm, sorghum midge, sowbugs, spittlebugs, spruce budworm, squash bug, stem beetles, stinkbugs, strawberry leaf roller, strawberry weevil, striped blister beetle, sunflower beetle, tadpole shrimp, tarnished plant bug, tentiform leaf miner, thornbug, three-cornered alfalfa hopper, thrips, ticks, tobacco budworm, tobacco fleabeetle, tomato fruitworm, tomato hornworm, Tussock moth, vegetable weevil, velvet bean caterpillar, wasps, webworms, Western bean cutworm, Western tussock moth, white apple leafhopper, willow leaf beetls, yellow scale.

CARBOFURAN, FURADAN   Mfg: FMC

Alfalfa blotch leaf miner, alfalfa snout beetle, alfalfa weevil, aphids, armyworms, banana root borer, banks grass mite, cereal leaf beetle, chinch bugs, clearwinged bug, clearwinged borer, Colorado potato beetle, cone beetles, cone worms, cone borer, cottonwood leaf beetle, cottonwoood twig borer, corn rootworms, cutworms, elm leaf beetle, European corm borer, flea beetles, grape phylloxera, grasshoppers, greenbugs, hornworms, lygus bugs, Mexican bean beetle, mosquitoes, nematodes, Pales weevil, pitch eating weevil, potato leafhopper, potato tuberworm, rice water weevil, root weevil, seed bugs, seed corn maggot, Southern corn rootworm, Southwestern corn borer, stalk rot, sugar beet maggots, sugarcane borer, sunflower beetle, sunflower weevil, thrips, tobacco budworm, tuberworm, tung borer, white grubs, wireworms.

CHECKMATE TPW (pheromone)   Mfg: Concep

Tomato pin worm.

CHINOMETHIONATE, MORESTAN   Mfg: Miles Inc.

Mites, pear psylla, powdery mildew, whiteflies.

CHLORPYRIFOS-METHYL, RELDAN   Mfg: DowElanco, Gustafson

Angoumoirs grain moth, confused flour beetle, flour beetles, granary weevils, Indian meal moth, lesser grain beetle, lesser grain borers, rice weevils, sawtoothed grain beetles.

CHLORPYRIFOS, DURSBAN, ESTATE,   Mfg: DowElanco
LORSBAN, PAGEANT, LOCK-ON, TENURE

Adelgibs, ants, aphids, apple maggot, armyworms, ash borer, bagworms, banded sunflower beetles, billbugs, bollworms, boll weevil, borers, boxelder bugs, brown dog ticks, carpet beetles, catalpa sphinx moth, centipedes, chiggers, chinch bugs, citrus rust mites, cockroaches, codling moth, cankerworms, corn borers, corn earworm, corn rootworm, cottonwood beetles, cotton leaf perforator, cranberry spanworm, cranberry weevil, crickets, cutworm, dogwood borer, earwigs, Eastern tent caterpillars, Egyptian alfalfa weevil, elm bark beetle, elm spanworm, European apple sawfly, European chafer, European cornborer, European crane fly, eyespotted budmoth, fall webworm, fiery skyger, fireants, firebrats, fireworms, fleabeetle, fleahopper, fleas, fruit leafroller, gnats, grasshoppers, green bugs, green fruitworm, hickory shuckworm, hornets, hyperodes weevil, imported cabbageworm, Indian meal moth, Japanese beetle, jackpine budworm, June beetle, Juniper webworm, katydids, keds, lacebugs, leafcutters, leafhoppers, leafminers, leafrollers, lesser appleworm, lesser cornstalk borer, lesser peach tree borer, lilac borer, lucerne moth, lygus, Mahogany webworm, maple leaf cutter, mealybugs, Mediterranean flour moth, millipedes, mimosa webworm, mint root borer, mirids, mites, mole crickets, moths, naval orangeworms, needle miners, oakworms, oak skeletonizer, oblique banded leafroller, oleander caterpillar, onion maggots, orange stripped oakworm, orange tortix, oriental fruit moth, pales beetles, pandomis leafroller, peach treeborer, peach twigborer, pecan nut casebearer, pecan weevil, periodical cicada, phylloiera, pink bollworm, pitch eating weevil, plant bugs, plum curculio, poplar tentmaker, puss caterpillar, red banded leafroller, red flour weevil, red humped caterpillar, rhododendron borer, rice weevils, root maggots, rose chafer, salt marsh caterpillar, San Jose scale, sawbugs, sawflies, sawtoothed grain beetles, scales, scorpions, silverfish, sod webworms, Southern mashed chafer, Southern pine beetles, Southwestern corn borer, spargnothis fruit worms, spiders, spittlebugs, spring elm caterpillars, springtails, spruce budworm, sugar beet root maggots, symphylons, tarnished plant bug, tent caterpillars, termites, thornbug, thrips, ticks, tobacco budworm, tufted apple budmoth, varigated leafroller, vegetable webworms, walnut caterpillar, walnut scale, wasps, webworms, weevils, Western tussock moth, whiteflies, white grubs, wireworms, wolly bears, yellow jackets, yellow-necked caterpillar.

CLOFENTEZINE, APOLLO　　　　　　　　　Mfg: Agr Evo

Mites.

CODLURE (pheromone),　　　　　　　　　Mfg: Biocontrol, Concep
ISOMATE-C, CHECKMATE-CM

Codling moth.

COUMAPHOS, CO-RAL　　　　　　　　　Mfg: Miles Inc.

Fleeceworms, grubs, hornflies, keds, lice, screwworms, ticks, face flies, scab mite, wool maggots.

CYFLUTHRIN, BAYTHROID, TEMPO　　　　　Mfg: Miles Inc.

Ants, aphids, armyworms, azalea caterpillar, bagworms, bed bugs, bees, billbug, black vine weevil, bluegrass, boll weevil, bollworms, boxelder bugs, budworms, cabbage looper, Calfiornia oakworms, canker worms, carpet beetles, casebearers, centipedes, chiggers, chinch bug, crickets, clothes mith, clover mites, cockroaches, cotton aphids, cotton perforator, crickets, cutworms, earwigs, elmleaf beetle, elm spanworm, European corn borer, firebrats, fleas, fleahopper, fleas beetles, flies, gnats, grasshoppers, gypsy moth, hornets, hyperodes weevil, Japanese beetle, June beetles, lace bugs, leaf hoppers, leaf rollers, leaf skeletonizers, leaf worms, loopers, lygus, mealy bugs, midges, millipedes, mole crickets, mosquitos, oleander moth, pear psylla, pillbugs, pine shoot moth, pink bollworms, plant bugs, red humped caterpillar, rose slugs, sawflies, scale, scorpions, silverfish, sow bugs, soybean looper, spiders, spittle bugs, striped beetles, striped oakworm, tent caterpillar, thrips, ticks, tip moth, tobacco budworm, Tussock moth, wasps, webworms, whiteflies, yellow necked caterpillars, leafminers, grasshoppers, salt marsh caterpiller, stinkbug, 3-cornered alfalfa hopper.

CYPERMETHRIN, AMMO, DEMON,　　　　　Mfg: Zeneca and FMC
CYMBUSH, CYNOFF

Alfalfa looper, ants, armyworm, beet armyworm, bees, black pecan aphid, boll weevil, cabbage looper, centipedes, cockroaches, corn borer, corn earworm, cotton, bollworm, cotton leafhopper, cotton leaf perforator, crickets, cutworms, fall armyworms, firebrats, fleas, flies, hickory shuckworm, lygus bugs-plant bugs, millipedes, pecan nut casebearer, pecan weevil, pillbugs, pink bollworm, silverfish, sowbugs, soybean thrips, spiders, tarnished plant bugs, ticks, tobacco budworm, tobacco thrips, wasps, whiteflies, yellow pecan aphid, yellow striped armyworm, stinkbugs, budworms, cabbagewom, bees, box elder bugs, earwigs, saltmarsh caterpiller.

CYROMAZINE, LARVADEX,   **Mfg: CIBA**
TRIGARD, CITATION, ARMOR

Flies, leafminers, sciarid flies.

DAZOMET, BASAMID   **Mfg: BASF**

Weeds, nematodes, soil borne diseases, soil borne insects.

DELTAMETHRIN, K-OTHRINE,   **Mfg: Roussel Bio/**
SUSPEND   **DowElanco**

Cockroaches, crawling insects.

DIAZINON, KNOX-OUT   **Mfg: CIBA, Elf-Atochem,**
          and others

Alfalfa weevil, ants, apple aphid, apple maggot, bagworms, banded cucumber beetles, bean aphid, Bermuda mite, billbugs, black cherry aphid, blackheaded fireworms, black scale, brown soft scale, cabbage maggot, carnation bud mite, carpet beetles, carrot rust flies, cherry fruit flies, cherry rust mites, chiggers, chinch bug, citrus aphid, clover mite, cockroaches, codling moth, Colorado potato beetle, corn earworm, corn rootworm, corn sap beetle, cottony cushion scale, cotton leaf perforator, cotton leafworm, cranberry fruitworm, cranberry girdler, crickets, cutworms, cyclamen mite, diamondback moth, dipterous leaf miner, earwigs, European fruit lecanium, European red mite, eye-spotted budmoth, fall armyworm, filbert leaf roller, flea beetles, fleas, flies, Forbes scale, fruit treeleaf roller, grape berrymoth, grape leaffolder, grasshoppers, green apple aphid, Harlequin bug, holly budmoth, houseflies, imported cabbageworm, keds, lawn billbug, leaf curl plum aphid, Mexican bean beetle, Mimosa webworm, mites, olivescale, onion thrip, onion maggot, Oriental fruit moth, Pacific spider mite, Parlotoria scale, pea aphid, peach twigborer, pear leaf blister mite, pear psylla, pecan nut casebearers, plant bug, privet mite, raspberry rootborer, rosy apple aphid, San Jose scale, sciarids, scorpions, seedcorn maggot, serpentine leafminer, silverfish, sod webworm, sorghum midge, Southern armyworm, Southern corn rootworm, Southwestern cornborer, sowbugs, spiders, spittlebug, spotted alfalfa aphid, springtails, strawberry leaf roller, tentiform leafminer, thrips, ticks, two-spotted mite, vinegar flies, walnut caterpillars, walnut scale, whiteflies, white grubs, wireworms, hyperodes weevil, woolly apple aphid, yellow clover aphid, yellow jackets.

DICHLOROVOS, DDVP, VAPONA   Mfg: Amvac

Ants, aphids, bedbugs, brown dog ticks, carpet beetle, centipedes, cigarette beetles, clothes moths, clover mites, cockroaches, crickets, dried fruit beetle, fleas, flies, flying moths, fruit flies, gnats, hornets, lice, mealybugs, mites, millipedes, mosquitoes, phorid flies, roaches, sciarid flies, scorpions, silverfish, sowbugs, spiders, ticks, tobacco moths, wasps, whiteflies.

DICHLOROPROPANE, TELONE II   Mfg: DowElanco

Bacterial canker, nematodes, soil insects, soil rot, verticillium wilt.

DICOFOL, KELTHANE   Mfg: Rohm & Haas

Almond mite, apple rust mite, brown mite, carmine mite, citrus flat mite, citrus red mite, citrus rust mite, clover mite, desert mite, European red mite, grass mite, McDaniel mite, Pacific mite, peach silver mite, plum nursery mite, schoene mite, six-spotted mite, strawberry mite, tomato russet mite, tropical russet mite, two-spotted mite, Williamette mite, yellow mite, Yuma mite.

DICROTOPHOS, BIDRIN   Mfg: Amvac

Aphids, thrips, mites, fleahoppers, grasshoppers, boll weevils, stinkbugs, lygus bugs, saltmarsh caterpiller, leaf perforators.

DIENOCHLOR, PENTAC   Mfg: Sandoz

Mites and whiteflies.

DIFLUBENZURON, DIMILIN   Mfg: Uniroyal

Armyworms, boll weevil, codling moth, Douglas fir tussock moth, forest tent caterpillars, green cloverworm, Gypsy moth, Mexican bean beetle, mosquitoes, Nantucket pine tipmoth, range caterpillar, sciarid fly, spruce budworm, terminal weevil, velvet-bean caterpillar.

DIMETHOATE, CYGON   Mfg: Numerous

Alfalfa hopper, alfalfa weevils, aphids, bagworms, banks grassmite, beanbeetle, bean leafbeetle, corn rootworms, Douglas fir conemidge, European pine shootmoth, flea hoppers, grape leafhoppers, grasshoppers, houseflies, Iris borer, lace bugs, leafhoppers, leaf miners, Loblolly pine sawfly, lygus bugs, Mexican bean beetle, midges, mites, Nantucket pine tip moth, pear psylla, pepper maggot, plantbugs, scales, thrips, whiteflies, Zimmerman pine moth.

DISULFOTON, DI-SYSTON   Mfg: Miles Inc.

Aphids, banks grassmite, birch treeminer, brown soft scale, camellia scale, Colorado potato beetle, elm leafbeetle, European elm scale, fleabeetle, grasshoppers, Hessian fly, holly leaf miner, lace bug, leafhoppers, leaf miners, mealybug, Mexican bean beetle, Mimosa webworm, mites, pine tip moth, potato psylid, root maggot, rootworms, scales, sorghum midge, Southern potato wireworm, thrips, whiteflies.

ENDOSULFAN, THIODAN, PHASER   Mfg: FMC & Agr Evo

Aphids, army cutworm, armyworm, artichoke plume moth, banded cucumber beetle, bark beetles, bean leaf beetle, bean leaf skeletonizer, black vine weevil, blister beetle, boll weevil, bollworm, boxelder bug, cabbage looper, catfacing insects, cereal leaf beetle, Colorado potato beetle, consperse stinkbug, corn earworm, cotton leaf perforator, cowpea curculio, cross-striped cabbageworm, cucumber beetles, cutworms, diamondback moth larvae, dogwood borer, European corn borer, eyespotted budmoth, false chinch bug, filbert leafroller, fleabeetles, fruit tree leafroller, garden symphylan, grape leafhopper, green June bug larvae, green stinkbug, Harlequin bug, hornworms, imported cabbageworm, iris borer, leaf-footed bug, leaf miner, letherleaf fern borer, lesser peach tree borer, lilac borer, lygus bugs, meadow spittlebug, melon leafhopper, melon leaf miner, melonworm, Mexican beanbeetle, mites, omnivorous leafroller, pea weevil, peach treeborer, peach twigborer, pear psylla, pecan nut casebearer, pillbugs, plant bugs, potato fleabeetles, potato leafhopper, potato psyllid, potato tuberworm, rindworm, rose chafer, serpentine leafminer, Southern green stinkbug, squash beetle, squash bug, squash vineborer, stinkbugs, striped fleabeetle, sugar beet webworm, sugarcane borer, sunflower moth, sweet potato fleabeetle, tarnished plantbug, three-lined potato beetle, thrips, tobacco budworm, tobacco fleabeetle, tobacco hornworm, tomato fruitworm, tomato hornworm, webworm, Western bean cutworm, whiteflies, wood borers, yellow-striped armyworm, Zimmerman pine moth.

ENZONE Mfg: Unocal

Phylloxera, nematodes, oakroot fungus, phytophthora rootrot.

ESFENVALERATE, ASANA, Mfg: DuPont
CONQUER

Alfalfa caterpiller, alfalfa looper, American plum borer, apple aphid, apple budmoth, apple maggot, artichoke plumemoth, aster leafhopper, balsam twig aphid, balsam wooly adelgid, banded cucumber beetle, banded sunflower moth, beanleaf beetle, beet armyworm, black cherry aphid, boll weevil, buckthorn aphid, cabbage looper, carrot weevil, celery looper, cherry fruitfly, chinch bugs, codling moth, Colorado potato beetle, coneworms, corn earworm, corn leaf aphids, corn rootworm, cotton bollworm, cotton leafworm, cowpea curculio, cranberry gridler, cucumber beetle, cutworm, European cornborer, fall armyworm, filbert worm, fleabeetles, fleahoppers, grasshoppers, green cloverworm, green fruit-worm, Kaliothis, hickory shuckworm, imported cabbageworm, Japanese beetle, leafhoppers, lesser apple worm, lesser cornstalk borer, lesser peach tree borer, lygus, Mexican beanbeetle, Nantucket pine tip moth, navel orangeworm, Northern pine weevil, oat-bud cherry aphid, oblique banded leafroller, oriental fruit moth, painted lady butterfly, pales weevil, pea aphid, peach tree borer, peach twig borer, pear psylla, pear slug, pecan sphids, pecan leaf phylloxera, pecan nut casebearer, pecan spittlebug, pecan stem phylloxera, pecan weevil, pepper weevil, periodi-cal cicada, pickleworm, pine chafer, pine conelet bug, pine leaf chemid, pine needle midge, plant bugs, plum curculio, potato leafhopper potato psyllid, potato tuberworm, red banded leafroller, red pine sawfly, red-necked peanut worm, redheaded pine sawfly, rindworm, rosy apple aphid, saltmarsh caterpillar, San Jose scale, sap beetles, seed chalid, seedbugs, Southern green stinkbug, Southwestern corn borer, spittle-bugs, spruce budworm, squash bug, squashvine borer, stalk borer, stinkbugs, sugarcane borer, sunflower maggot, sunflower moth, sun-flower stem weevil, sunflower seed weevil, tentiformleafminer, three cornered alfalfahopper, thrips, tobacco budworm, tobacco hornworm, tomato fruitworm, tomato hornworm, tomato pinworm, tufted apple bud-worm, upland bugs, varigated leafroller vegetable leafminer, velvetbean caterpiller, walnut aphid, walnut huskfly, Western bean cutworm, white apple leafroller, whitefly, woolybear caterpiller.

ETHION Mfg: FMC

Aphids, chinch bug, cucumber beetles, leafhoppers, leafminer, melon leafminer, mites, scales, serpentine leafminer, sod webworm, white flies.

ETHOPROPHOS, MOCAP — Mfg: Rhone Poulenc

Black turfgrass ateanius, bluegrass billbug, cinch bugs, corn rootworms, cutworms, European chafer, fleabeetles, garden symphylans, Japanese beetle, nematodes, sod webworms, wireworms.

EXHIBIT (beneficial nematodes) — Mfg: Biosys

Black vine weevil, strawberry root weevil, fungus gnats, white grubs, cutworms, sod webworm, billbugs.

FENAMIPHOS, NEMACUR — Mfg: Miles Inc.

Nematodes, bacterial canker, thrips, and mole crickets.

FENBUTATIN- OXIDE, VENDEX — Mfg: DuPont

Mites.

FENOTHRIN, SUMITHRIN — Mfg: Olympic

Whiteflies, aphids, mites, meely bugs.

FENOXYCARB, TORUS, AWARD, PRECISION, ECLIPSE — Mfg: Ciba

Cockroaches, fleas, fireants, leafminers, whiteflies, scales, fungus gnats, shoreflies.

FENPROPATHRIN, TAME, DANITOL — Mfg: Valent

Leafhoppers, whitefly, mites, beet armyworm, mealybug, aphids, lacebugs, leafhoppers, leafminers, Japanese beetle, scales, armyworms, pink bollworm, whiteflies.

FENTHION, BAYTEX — Mfg: Miles Inc.

Mosquitoes.

FENVALERATE, PYRID, ECTRIN, TRIBUTE — Mfg: DuPont and others

Alfalfa looper, ants, aphids, apple maggot, armyworm, artichoke moth, banded sunflower moth, bean leafbeetle, black cutworm, boll weevil, cabbage looper, cadelles, carpetbeetles, carrot weevil, centipedes, cherry fruitfly, chinch bugs, codling moth, cigarette beetles, cockroaches, Colorado potato beetle, conebugs, confused flour beetle, corn earworm, corn rootworm, cotton bollworm, cotton leaf perforator, crickets, cucumber beetle, cutworms, cowpea curculio, diamondback moth, drug store beetle, European cornborer, fleabeetles, fleas, flies, leafminers, fruitworms, granary weevil, grasshoppers, green cloverworm, Heliothis Spp., hickory shuckworm, hornworm, imported cabbage worm, leafhoppers, leaf miners, leaf rollers, lesser appleworm, lesser peach treeborer, lice, lygus, mealmoth, Mexican bean beetle, Nantucket pine tipmoth, Naval orangeworm, Oriental fruit moth, pealeaf weevil, peach twigborer, pear psylla, pear rust mite, pear slug, pecan casebearer, pecan weevil, periodical cicadas, pickleworm, pinworm, pink bollworm, plum borer, plum curculio, potato psyllid, potato tuberworm, red blanket leafroller, red-necked peanutworm, rindworm, red flour beetle, saltmarsh caterpillar, San Jose scale, saw-toothed grain beetle, scales, seed bugs, seedweevil, sheep keds, silverfish, Southern green stinkbug, Southwestern cornborer, sowbugs, spiders, squash bug, sugarcane borer, sunflower moth, sunflower stem weevil, tarnished plant bug, tent, ticks, tobacco budworm, tufted apple budworm, vegetable leafminer, Velvetbean caterpillar, weevils, Western bean cutworm, whiteflies.

FLUVALINATE, MAVRIK, YARDEX   Mfg: Sandoz

Ants, aphids, armyworms, boll weevils, bollworm, Cabbage seed pod weevil, caterpillars, chiggers, chinchbug, clover mites, corn earworm, cotton leaf per-
forator, crickets, cucumber beetle, cutworm, diamond back moth, earwigs, elmleaf beetles, flea beetles, fleahoppers, fleas, grasshoppers, greenscale, gypsy moth, imported cabbageworm, Japanese beetle, leaf feeding caterpillars, leaffolders, leafhoppers, leaf miners, leafrollers, loopers, lygus, mealy bugs, milipedes, mites, oriental fruit moth, peach twig borer, pear psylla, root weevils, saltmarsh caterpillar, scales, sowbugs, springtails, stinkbugs, tentiform caterpillar, thrips, ticks, tobacco budworm, webworms, weevils, whiteflies.

FONOFOS, DYFONATE, CRUSADE, MAINSTAY   Mfg: Zeneca

Aphids, billbugs, cabbage maggot, chinch bugs, corn rootworms, cutworm, earwigs, European corn borer, garden symphylans, Gelechiid moth, greenbugs, lesser cornstalk borer, mole crickets, onion maggot, peanut rootworm, seed corn beetle, seedcorn maggot, sod webworm, sugarbeet root maggot, symphylans, white grub, wireworms.

FORMETANATE, CARZOL   Mfg: Agr Evo

Apopka weevil, consperse stinkbug, lygus bug, mites, tentiform leaf miners, thrips, citrus root weevil, Fuller rose beetle, little leaf notcher.

GBM ROPES (pheromone)   Mfg: Biocontrol

Grape berry moth.

GOSSYPLURE (pheromone),   Mfg: Biosys,
PBW ROPE,DECOY,   Concep, Ecogen,
CHECKMATE, MEC   Mitsubishi & others

Pink bollworm.

HEXAFLUMURON, NAF-46   Mfg: DowElanco

Termites.

HEXYTHIAZOX, SAVEY   Mfg: Gowen

Mites.

HYDRAMETHYLNON, AMDRO,   Mfg: American Cyanamid
MAX FORCE, SEIGE

Imported fire ant, cockroaches.

HYDROPRENE, GENCOR, GENTROL   Mfg: Sandoz/Zoecon

Cockroaches.

IMIDACLOPRID, MERIT, ADMIRE,   Mfg: Miles Inc.
PROVADO, MARATHON

Aphids, fleabeetles, Japanese beetle, black turfgrass ataenius, masked chafer, European chafer, oriental beetle, asiatic garden beetle, phyllophaga spp., billbugs, annual bluegrass weevil, adelgids, aphids, elm leaf beetle, lacebugs, leafhoppers, leafminers, mealy bugs, sawfly larvae, thrips, white flies, white grubs, scale, pine tip moth, white grubs, billbugs, bluegrass weevils aphids, adelgids, lacebugs, leafminers, leaf feeding beetles, mealybugs, pine tip moth, scales, thrips, whiteflies, Colorado potato beetle.

**INSECTICIDAL SOAP, SAFER,
M-PEDE**

Mfg: Mycogen & Ringer

Adelgids, aphids, earwigs, grasshoppers, harlequin bugs, lacebugs, leafhoppers, lygus bugs, mealybugs, mites, pear psylla, pear slugs, plantbugs, psyllids, sawflys, scales, tent caterpillars, thrips, whitefly.

**ISOMATE-M (pheromone),
CHECKMATE OFM**

Mfg: Biocontrol, Concep

Oriental fruit moth.

ISAZOFOS, TRIUMPH

Mfg: CIBA

Annual bluegrass weevil, annual white grub, ants, armyworms, asiatic garden beetle, billbugs, bluegrass weevel, chinch bugs, clover mites, crickets, cutworms, dung beetle European chafer, grasshoppers, Hyperodes weevil, Japanese beetle, June beetle, leafhoppers, millipedes, mole crickets, Southern masked chafer, webworms.

ISOFENPHOS, OFTANOL

Mfg: Miles Inc.

Armyworms, Asiatic garden beetle, ataenius, billbugs, chafers, chinch bugs, corn rootworm, craneflies, clover mites, crickets, cutworms, fire ants, fleabeetle, green June beetle, grasshoppers, Hyperodes weevil, Japanese beetle, leafhoppers, millipedes, mole crickets, Oriental beetle, sod webworm, white grubs.

KINOPRENE, ENSTAR

Mfg: Sandoz/Zoecon

Whiteflies, aphids, scales, mealey bugs, fungus gnats.

LAGENIDIUM GIGANTEUM

Mfg: Calif. M.A.D.

Mosquitoes.

LAMBDA-CYHALOTHRIN, KARATE,
COMMODORE, SCIMITAR

Mfg: Zeneca

Ants, armyworms, boll weevil, cabbage looper, carpet beetles, centipedes, cigarette beetle, crickets, cockroaches, confused flour beetle, cotton bollworm, cotton leafhopper, cotton leaf perforator, cutworms, European cornborer, firebrats, lesser grain borer, lygus, millipedes, pillbugs, pink bollworm, red flour beetles, rice weevils, saltmarsh caterpillars, saw toothed grain beetle, scorpions, silverfish, sowbugs, spiders, stink bugs, thrips, tobacco budworm, bagworm, blackvine weevil, caterpillars, aphids, boxelder bugs, budworms, oakworms, cankerworms, clover mites, crickets, Eastern tent caterpillar, elmleaf beetle, webworms, fleabeetles, gypsy moth, Japanese beetles, June beetles, lacebugs, leafhoppers, leaf rolls, leaf skeletonizers, mealybugs, midges, mosquitos, oleander moth, pine sawfly, plant bugs, root weevil, mites, spittlebugs, striped beetles, tipmoth, Tussock moth, wasps, whiteflies, armyworms, earwigs, grasshoppers, ticks, billbugs, chinch bugs, chiggers, Hyperodes weevil, mole crickets.

LINALOOL, DEMIZE   Mfg: Pet Chemical Co.

Fleas.

LINDANE   Mfg: Numerous

Ants, aphids, apple grain aphid, armyworms, bean weevil, bedbugs, borers, cabbage maggots, carpetbugs, chiggers, climbing cutworms, clothes moth, confused flourbeetle, crickets, cucumber beetle, curculio, Diabrotica, European sawfly, false wireworms, fleabeetles, fleas, fleeceworms, fowl tick, Fuller rose weevil, garden symphyhids, gladiolus thrip, gnats, granary weevil, grasshopper, houseflies, Indian meal moth, keds, lace bugs, leaf miners, lice, locust borer, loopers, lygus bugs, mange, melonworms, mosquitoes, mushroom flies, onion maggot, pecan curculio, pecan phylloxera, phylloxera, pickleworms, pine bark aphids, pine root collarbeetle, powderpost beetle, psyllid, rice weevil, roaches, rose midge, rosy apple aphid, saw-toothed grain beetle, screwworm, seed corn maggot, silverfish, spiders, spittlebugs, spruce gall aphid, aquash vineborer, stableflies, strawberry weevil, tarnished plant bug, termites, thrips, ticks, webworms, whitefly, white grubs, whitepine weevil, wireworms, woolly apple aphids.

LYCOLURE, NOMATE TPW, DECOY   Mfg: Ecogen, Biosys,
TPW, ISOMATE TPW   Biocontrol

Tomato pinworm.

MAGNESIUM PHOSPHIDE,   Mfg: Degesch and
FUNI-CEL, MAGTOXIN   Pestcon Systems

Almond moth, Angoumois grain moth, bean weevil, cadelle, cereal leaf beetle, cigarette beetle, confused flour beetle, dernested beetle, dried fruit beetle, dried fruit moth, European grain moth, flat grain beetle, fruit flies, granary weevil, Indian meal moth, lesser grain borer, Mediterranean flour moth, pink bollworm, raisin moth, red flour beetle, rice weevil, rusty grain beetle, saw-toothed grain beetle, tobacco moth.

MALATHION, CYTHION, FYFANON   Mfg: Cheminova

Alfalfa weevil, aphids, armyworms, bagworms, beet leafhopper, black cherry aphid, blackheaded fireworm, black scale, blueberry maggot, boll weevil, budmoth, cabbage looper, cereal leafbeetle, cherry fruitfly, chicken red mite, clover mite, codling moth, confused flour beetle, corn rootworm, cranberry fruitworm, cucumber beetle, curculio, drosophila, European red mite, false chinch bug, flat grainbeetle, field chickets, flies, Florida red scale, Forbes scale, fruit tree leafroller, granary weevil, grasshoppers, green apple aphid, hornflies, imported cabbageworm, Indian meal moth, Japanese beetle, juniper scale, ked, lacebugs, leafhoppers, lesser grain borer, lice, lygus, mealybugs, mites, Monterey pine scale, mosquitoes, Northern fowl mite, onion maggot, Oriental fruit moth, peach treeborer, pear psylla, pickleworm, pine needle scale, plum curculio, potato aphid, potato leafhopper, poultry lice, poultry ticks, purple scale, red-banded leafroller, red flour beetle, rice stinkbug, rice weevil, rose scale, rosy apple aphid, rusty grain beetle, saw-toothed grain beetle, soft brown scale, soft scale, spittlebug, squash vineborer, strawberry leafroller, strawberry root weevil, tarnished plantbugs, tent caterpillars, thrips, tick, two-spotted mites, unspotted tentiform leafminer, vetch bruchids, whiteflies, Williamette mite, woolly apple aphid, yellow-backed caterpillars, yellow scale.

METALDEHYDE, DEADLINE   Mfg: Numerous

Slugs and snails.

METARHIZIUM ANISOPILAE, BIOPATH   Mfg: Eco Science

Cockroaches, fleas, termites, whiteflies, aphids.

METHAMIDOPHOS, MONITOR   Mfg: Miles Inc. and Valent

Aphids, beet armyworm, cabbage looper, Colorado potato beetle, cutworms, diamond back moth, flea beetles, imported cabbageworm, leafhoppers, mites, potato tuberworm, stinkbugs, thrips.

METHAM-SODIUM, VAPAM,   Mfg: Zeneca and others
SOIL PREP, SECTAGON II

Nematodes, symphylids, soilborne diseases such as Rhizoctonia, Pythium, Phytophtora, Verticillium, Sclerotinia, oak root fungus, club root, Dutch elm disease, weed seeds, soil insects.

METHIDATHION, SUPRACIDE   Mfg: CIBA

Alfalfa weevil, aphids, boll weevil, banksgrass mite, bollworms, budworms, citrus blackfly, codling moth, Egyptian alfalfa weevil, fleabeetles,
greenbug, hickory shuckworm, hornworms, leafhoppers, leafworms,
lygus, peach twigborer, scales, spider mites, spittlebugs, whiteflies.

METHOMYL, LANNATE   Mfg: DuPont

Alfalfa caterpillar, alfalfa looper, alfalfa weevil, aphids, appleworms,
armyworms, beet webworm, bollworm, cabbage looper, cabbageworm,
codling moth, corn earworm, cotton leaf perforator, cutworms, diamondback moth, European cornborer, fall armyworm, fleabeetles, grape berry
moth, grape leafholder, grape leafrollers, green cloverworm, green
fruitworm, imported cabbageworm, leafhoppers, leafminers, lygus bugs,
melonworms, Mexican bean beetle, omnivorous leafroller, orange tortrix,
Oriental fruit moth, picnicbeetle, pickleworm, pinworms, potato tuberworm, twigborers, velvetbean caterpillar, tufted apple budmoth, fruit tree
leafroller, spruce budworm, sorghum midge, Western tussock moth,
orange tortrix, leaf perforator, asparagus beetle, tarnished plant bug.

METHOPRENE, ALTOSID, APEX,   Mfg: Sandoz/Zoecon
DIANEX, MINEX, DIACON,
PRECOR, PHARORID, PETCOR

Almond moth, ants, cigarette beetles, confused flour beetle, fleas, flies,
Indian meal moths, leafminers, lesser grainborer, merchant grain beetle,
mosquitoes, Pharoah's ant, red flour beetle, saw-toothed grainborer,
Sciarid beetles, tobacco moth.

METHOXYCHLOR, MARLATE   Mfg: Kincaid Mfg.

Alfalfa caterpillar, alfalfa weevil, apple maggot, armyworms, asparagus
beetle, bean leaf beetle, blister beetle, cabbageworm, Cadella flat
grainbeetle, cankerworm, caterpillar, cherry fruitfly, cherry fruitworm,
codling moth, clover leaf weevil, Colorado potato beetle, confused flour
beetle, corn earworm, cowpea curculio, cranberry fruitworm, cucumber
beetle, fall armyworm, flat grain beetle, flea beetle, fleas, foreign grain
beetle, gnats, grape berry moth, grapeleaf skeletonizer, horn flies,
houseflies, Japanese beetle, keds, leafhoppers, lesser grain borer, lice,
long-headed beetle, melonworm, Mexican bean beetle, mosquitoes,
pear slug, pea weevil, plum curculio, red flour beetle, rice weevil, roaches,
rose chafer, rose slugs, San Jose scale, saw-toothed grain beetle,
soybean caterpillar, spittlebug, squash vineborer, stable flies, strawberry
weevil, tent caterpillars, ticks, velvetbean caterpillar, wasps, webworm.

METHYL BROMIDE   Mfg: Great Lakes Chem. Corp. and others

Stored grain insects, nematodes, commodity storage insects, fungi and weeds.

METHYL PARATHION, PENNCAP-M   Mfg: Numerous

Alfalfa caterpillar, alfalfa seed chalcid, alfalfa weevil, aphids, armyworms, artichoke plume moth, bean leafbeetle, black grassbug, blister beetle, boll weevil, bollworm, chinch bug, clover leaf weevil, clover seed chalcid, codling moth, corn earworm, corn rootworm, cotton leafworm, cotton leaf perforator cowpea curculio, cutworms (climbing), Egyptian alfalfa weevil, European pine shoot moth, false chinchbug, flea beetles, fleahoppers, grape leaf folders, grasshoppers, green clover worm, green June beetle, imported cabbageworm, leafhoppers, leaf miners, leaf rollers. loopers (cabbage), lygus bugs, Mexican bean beetle, mites, mosquitoes (larvae), Nantucket pine tipmoth, Oriental fruit moth, peach treeborer, plant bugs, plum curculio, potato psyllid, rice leaf miners, rice stinkbug, saltmarsh caterpillar, scales, seed corn maggot, sorghum midge, stinkbug, tadpole shrimp, three-cornered alfalfa hopper, thrips, velvetbean caterpillar, vetch bruchid, webworms.

NALED, DIBROM, LEGION   Mfg: Valent

Alfalfa webworms, aphid, armyworms, blister beetles, bollworms, cabbage looper, California oakworm, cankerworms, citrus cutworm, clover mites, cockroaches, cotton leaf perforator, crane flies, cucumber beetle, diamondback moth, drosophila, earwigs, elm leaf beetle, fleas, fleabeetle, flying moths, fruitflies, gnats, grasshoppers, hemlock borer worms, houseflies, imported cabbageworms, leafhoppers, leafminers, leaf perforator, leafroller, loopers, lygus bugs, mealybugs, midges, mosquitoes, moths, mites, oak webworm, omnivorous leaf tier, orange tortrix, Oriental fruit moth, peach twigborer, pickleworms, range caterpillar, red spider mite, saltmarsh caterpillar, sap beetle, strawberry leafroller, Sierra fir borer, soft brown scale crawlers, spider mites, spittlebugs, stable flies, stinkbugs, tent caterpillars, thrips, ticks, tomato fruitflies, Tussock moth, webworms, whiteflies, Zimmerman pine moth, grape leaf skeletonizer.

NOSEMA LOCTUSTAE,   Mfg: M&R Durango & others
GRASSHOPPER SPORE,
NOLO BAIT

Grasshoppers, mormon crickets.

OXAMYL, VYDATE   Mfg: DuPont

Aphids, blackvine weevil, boll weevil, Colorado potato beetle, cottonleaf perforator, fleabeetles, fleahopper, fungus fly, fungus gnats, green peach aphid, Japanese beetle, leafhoppers, leafminers, mealybugs, mites, nematodes, pepper weevil, Royal palm bug, scales, thrips, whitefly.

PARATHION   Mfg: Numerous

Alfalfa caterpillar, alfalfa seed chalcid, alfalfa weevil, alfalfa weevil larvae, American cockroach, American plum borer, ants, aphis, apple flea weevil, apple lace bug, apple maggot, apple mealybug, apple redbugs, armyworms, artichoke plume moth, Asiatic garden beetle, avocado lacebug, avocado leafhopper, bagworms, bean leafbeetle, beet crown borer, black grass bug, black vine weevil, blister beetle, blossom anomala, blossom weevil, blueberry maggot, blueberry tipworm, boll weevil, boll-worms, budworm, cabbage seedpod weevil, cankerworms, carrot rust fly, casebearer, cat facing insects, celery leaftiers, cherry fruitworm, chinch bugs, citrus root weevil, climbing cutworms, clover head weevil, clover leaf weevils, codling moth, Colorado potato beetle, consperse stinkbug, corn earworms, corn rootworms, corn silkfly, cotton leafperforators, cotton leafworms, crane flies (larvae), cranberry fruitworm, cranberry tipworm, crickets, crown borer, cucumber beetle, cucumber beetle (banded), currant borer, cutworms, darkling ground beetles, diamond-back moth, earwigs, European apple sawfly, European cornborer, European fruit lecanium scale, eye-spotted bud moth, false celery leaf tier, false chinch bug, fireworm, blackheaded flea beetles, fleahoppers, fruit flies, fruitworms, fuller rose beetle, grape berry moth, grape budbeetle, grasshoppers, green cloverworm, green fruitworm, green June beetle, greenhouse leaf tier, Harlequin bug, hoplia beetle, hornworms, imported cabbageworm, Japanese beetle, Katydids, lace bugs, leaf beetles, leaf folders, leaffooted bugs, leafhoppers, leaf miners, leafrollers, leaftiers, lesser appleworm, lesser peach tree borer, lesser cornstalk borer, lima bean pod borer, little fire ant, loopers, mealybugs, melonworm, Mexican bean beetle, midges, mites, mosquitoes, obscure weevil, onion maggot, orange tortrix, orangedog, orangeworms, Oriental fruitmoth, Pameras, pandemis moth, pea moth, pea weevil, pear psylla, peach bark beetle, peach treeborer, peach twigborer, pearbug, pecan leaf casebearer, pecan nut casebearer, pepper maggot, pickleworm, pink scavenger caterpillar, plant bugs, plum curculio, potato psyllid, potato tuberworm, psyllids, pumpkin bug, Quince curculio, raspberry crown borer, rednecked crane borer, rednecked peanutworm, rice leafminer, rindworm, rose chafer, salt marsh caterpillar, San Jose scale, sawflies, Say's stinkbug, scales seed corn maggot, shothole borer, snails (climbing), sorghum midge, sowbugs, spiders, spittlebugs, splitworm, spring tails, squash bugs, squash vineborer, stinkbugs, strawberry crown borer, strawberry rootworm, suckfly, Surinam cockroach, sweetclover weevil, symphylans, tadpole shrimp, tarnished plant bug, tent caterpillar, three-cornered alfalfa hopper, thrips, tip borers, tobacco budworm, tomato fruitworm, tomato pinworm, tortricid, moths, tortrix moths, tussock moths, twig girdlers, vegetable weevil, velvet caterpillar, walnut caterpillar, walnut husk fly, webworms, weevils, white grubs, whiteflies, wireworms, wood weevil.

PERMETHRIN, AMBUSH, ATROBAN,ECTIBAN, ASTRO POUNCE, PRAMEX, TORPEDO, PRELUDE

Mfg: FMC and Zeneca

Alfalfa caterpillar, alfalfa weevil, ants, apple aphids, armyworms, artichoke plume moth, asparagus beetle, bean leaf beetle, boll weevil, bollworms, budworms, cabbage looper, chinch bugs, cockroaches, codling moth, Colorado potato beetle, coneworms, corn earworms, corn borer, cotton aphids, cotton leaf perforator, crickets, crucifer weevil, cucumber beetles, cutworms, diamondback moth, European corn borer, eye gnats, filbertworm, flea beetles, fleahoppers, fowl mites, gnats, green cloverworm, green fruitworm, Heliothis Spp., hornworm, houseflies, imported cabbageworm, keds, leafhoppers, leafminers, leafrollers, lesser peachtree borer, lice hoppers, lygus, mange mites, melonworms, Mexican bean beetles, mosquitoes, mushroom flies, Nantucket pine top moth, naval orangeworm, oblique banded leafroller, Oriental fruit moth, peach twig borer, pear psylla, periodical circadas, pickleworm, pink bollworm, plum curculio, potato tuberworm, red banded leafroller, root weevils, rose chafer, scabies, seed bugs, sheep keds, soybean looper, spiders, squash bugs, squash vineborer, stable flies, stalk borers, stem weevils, stinkbugs, tarnished plant bug, tentiform leafminer, termites, thrips, ticks, tomato fruitworm, velvetbean caterpillars, walnut husk fly, webworms, whiteflies, asparagus beetle, bagworms, pine tip moth, lacebugs, coneworms, seed bugs, range caterpillar flies.

PETROLEUM OILS, DORMANT AND SUMMER OILS, SAF-T-SIDE, SUNSPRAY, VOLCK

Mfg: Numerous

Aphids, apple red bug, black scales, brown apricot scale, brown mite, bud mite, citricola scale, codling moth, European, fruit lecanium scales, European red mite, fig scale, frosted scale, fruit tree leaf roller, Italian pear scale, leaf roller, lichens, mealybugs, moss oleander scale, olive scale, Pacific mite, parlatoria, pear leaf blister mite, pear psylla, purple scale, Putnam scale, red scale, red spider mite, rust mite, San Jose scale, Scurfy scale, terrapin scale, twig borers, treehoppers, two-spotted mite, yellow scale, whiteflies, plant bugs, mites, scales, caterpillars, white flies, mealy bugs, aphids, leafminers, apple red bugs, fruit tree leaf roller, pear psylla, greasy spot, sooty mold psyllids, leaftiers, webworms, cankerworm, adelsids.

PHORATE, THIMET, RAMPART   Mfg: American Cyanamid

Alfalfa weevil, aphids, beet rootmaggot, black cutworm, clearwinged borers, Colorado potato beetle, cornleaf aphid, corn rootworm, cottonwood leaf beetle, cottonwood twig borer, European cornborer, fleabeetle, grasshoppers, leafbeetle, Hessian fly, leafhoppers, leafminers, lygus bugs, Mexican bean beetle, mites, Nantucket pine tipmoth, pea leafminers, planthoppers, psyllids, root lesion, nematodes, seedcorn maggots, seed corn beetles, Southern corn rootworm, Southern potato wireworm, spotted alfalfa aphid, sugarbeet root maggot, thrips, white grubs, whiteflies, wireworms.

PHOSMET, IMIDAN   Mfg: Gowan

Alfalfa weevil, aphids, apple maggot, birch leaf miner, blueberry maggots, boll weevil, cattle grubs, cherry fruitfly, codling moth, Colorado potato beetle, corn rootworm, cranberry fruitworms, Eastern tent caterpillar, elm spanworm, European pine shoot moth, green fruitworm, grubs, gypsy moth, grape berry moth, hornflies, Japanese beetle, leafhoppers, leaf rollers, lice, navel orangeworms, oblique banded leaf rollers, orange tortrix, Oriental fruit moth, pea leaf weevil, pea weevil, peach twig borer, pear psylla, pecan nut casebearer, plum curculio, potato fleabeetle, red humped caterpillars, scabies, scales, spring cankerworm, sweet potato weevil, tarnished plant bug, ticks.

PIRIMIPHOS-METHYL, ACTELLIC,   Mfg: Zeneca/Wilbur Ellis
NU GRO

Almond moth, angoremoris grain moth, cigarette beetle, confused flour beetle, corn sap beetle, flat grain beetle, granary weevil, hairy fungus beetle, Indian meal moth, maize weevil, merchant grain beetle, red flour beetle, rice weevil, sawtoothed grain beetle.

PROFENOFOS, CURACRON   Mfg: CIBA

Armyworms, boll weevil, bollworms, cabbage looper, cotton leaf perforator, mites, tobacco bud worms and whiteflies.

PROPARGITE, OMITE, COMITE, ORNAMITE

Mfg: Uniroyal

Mites.

PROPETAMPHOS, SAFROTIN

Mfg: Sandoz

Ants, box elder bugs, cigarette beetle, cockroaches, crickets, earwigs, fireants, fleas, grain beetle, ground beetle, red flour beetle, silverfish, spiders, ticks.

PROPOXUR, BAYGON

Mfg: Miles Inc.

Ants, billbugs, booklice, cadelle, carpet beetles, centipedes, chiggers, chinch bugs, cigarette beetle, clover mites, cockroaches, crickets, drugstore beetles, earwigs, fleas, fire brat, flies, flour moths, hornets, Indian meal moth, midges, millipedes, mole crickets, mosquitoes, punkies, sandflies, saw-toothed grain beetle, scorpions, silverfish, sod webworms, sowbugs, spiders, ticks, waterbugs, wasps, webworms.

PYRELLIN (pyrethrin/rotenone)

Mfg: Webb Wright

Aphids, asparagus beetle, alfalfa webworm, bean beetles, beet webworms, cabbage worms, celery leaf tiers, fleahoppers, harlequin bug, Japanese beetle, leafhoppers, leafminers, loopers, lygus bugs, mites, leafrollers, plantbugs, stinkbugs, thrips, whiteflies, armyworms , chinch bugs, flies, pink bollworm, fruit flies, mealybugs, Mexican bean beetle, blueberry maggots, cherry fruit flies, gooseberry fruit worms, imported current worms, red-necked borers, raspberry fruit worms, rose chafers, strawberry fruit worms, Fuller rose beetle, caterpillers, fireworms, weevils, Colorado potato beetle, cucumber beetle, squash bugs, European corn borer, fleabeetles, squash vine borer, vegetable weevils.

PYRENONE
(Combination of piperonyl butoxide & pyrethrin)

Mfg: Agr-Evo

Ants, aphids, asparagus beetle, beetles, blister beetles, boxelder bugs, cabbage looper, cabbageworm, Cadelles centipedes, cankerworms, cigarette beetles, Colorado potato beetle, crickets, cucumber beetle, diamondback moth, earwigs, firebrats, fireworms, fleabeetles, fleas, flies, flour beetles, fruit flies, gnats, grain moths, gypsy moth, Harlequin bugs, hornflies, hornets, horseflies, leafhoppers, leaftier, lice, med moths, medworms, Mexican bean beetle, mosquitoes, moths, plum moth, psoceds, roaches, silverfish, spiders, stable fleas, stinkbugs, stored grain insects, tarnished plant bug, vinegar flies, wasps, waterbugs, webworms, weevils, whiteflies.

PYRETHRIN   Mfg: Roussel Bio, MGK, Prentiss and others.

Ants, aphids, asparagus beetle, cabbage looper, cabbageworm, celery leaftier, celery looper, Colorado potato beetle, crickets, cucumber beetle, diamondback moth, fireworm, fleabeetles, flies, fruit flies, gnats, grain moths, Harlequin bug, Hawaiian beet webworm, leafhopper, leafroller, lice, Mexican bean beetle, midge, mosquitoes, psyllid, roaches, silverfish, sod webworms, spiders, thrips, ticks, twelve-spotted beetle, wasps, webworms.

RAYANOCIDE, RYANA   Mfg: Dunhill

Thrips, codling moth, European cornborer.

RESMETHRIN, SBP1382,   Mfg: Roussel Bio
SCOURGE, DERRINGER, VECTRIN

Ants, aphids, blackflies, centipedes, dog fleas, earwigs, fleas, flies, flour beetles, gnats, grain moths, hornets, japanese beetle, leafhoppers, mosquitoes, moths, plantbugs, roaches, silverfish, sowbugs, spiders, spittlebugs, thrips, wasps, whiteflies.

ROTENONE   Mfg: Roussel Bio, Uniroyal and others.

Aphids, Colorado potato beetle, fire ants, fish, leafrollers, orange tortrix, scales, spider mites.

SABADILLA, VERATRAN D   Mfg: Dunhill

Thrips.

SLAM, ADIOS (carbaryl/cucurbita   Mfg: Micro Flo/BASF
root powder)

Cucumber beetles, corn rootworms.

SODIUM FLUOALUMINATE, KRYOCIDE, CRYOLITE
Mfg: Elf-Atochem, Gowan

Armyworm, bean leafbeetle, bean leafroller, blister beetle, cabbage looper, cabbageworm, codling moth, corn earworm, cranberry weevil, cucumber beetle, cutworms, darkling ground beetle, diamondback moth, earwigs, fleabeetles, fleaweevil, fruitworm, Fuller rose beetle, garden tortrix, grape leaffolder, gypsy moth, hornworm, katydid, leafrollers, melonworm, Mexican bean beetle, orange dog, orange tortrix, pepper weevil, pickleworm, pinworm, plum curculio, rose weevil, strawberry root weevil, tobacco budworm, Western skeletonizer.

SPODOPTERA EXIQUA NPV, SPOD-X
Mfg: Crop Genetics

Beet armyworms.

SULFLURAMID, FLUORGUARD, VOLCANO
Mfg: Griffin/FMC

Ants, cockroaches.

SULFOTEPP, PLANTFUME
Mfg: Plant Products Corp.

Mites, thrips, scale, mealy bug, whiteflies.

SULFURYL FLORIDE, VIKANE
Mfg: DowElanco

Bedbugs, carpet beetles, clothes moths, cockroaches, deathwatch beetles, mice, old house borers, powder post beetles, termites, rodents.

SULPROFOS, BOLSTAR
Mfg: Miles Inc.

Armyworms, cotton bollworm, fleahoppers, green cloverworm, Heliothis spp., leafhoppers, lygus, Mexican bean beetle, pink bollworms, plant bugs, three cornered alfalfa hopper, thrips, tobacco budworm, velvetbean caterpillar.

TEFLUTHRIN, FORCE, FIREBAN
Mfg: Zeneca

Corn rootworms, cutworms, fireants, lesser cornstalk borer, seed corn maggot, wireworms, white grubs, seed corn beetle.

TELONE C-17 (dichloropropane/ chloropicrin)   Mfg: DowElanco

Nematodes, symphylans, wireworms and certain soil bourne diseases.

TEMEPHOS, **ABATE**   **Mfg: American Cyanamid**

Blackfly, buffalo gnat, mosquitoes, midges, gnats, punkies, sand flies.

TENNAX (fonofos/phorate)   **Mfg: Platte Chemical**

Wireworms, aphids, Colorado, potato beetle, flea beetle, leaf hoppers, leaf miners.

TERBUFOS, **COUNTER**   **Mfg: American Cyanamid**

Billbugs, corn rootworms, nematodes, seed corn beetle, seed corn maggots, symphylans, white grubs, wireworms, greenbugs, root maggots, leafhoppers, aphids, chinch bugs, fleabeetles, thrip, European corn borer.

TETRACHLORVINPHOS, **RABON**   **Mfg: Fermenta**

Flies, lice.

TETRALATE (tetramethrin/ resmethrin)   **Mfg: Roussel Bio**

Cockroaches, ants, spiders, silverfish, crickets, centipedes, earwigs, sowbugs, grain mites, spider mites, white flies, houseflies, mosquitos, gnats, flying moths, gypsy moth, tent caterpillers, cankerworms, garden webworms.

THIODICARB, **LARVIN**   **Mfg: Rhone Poulenc**

Armyworms, bean leaf beetle, boll weevil, bollworm, cabbage looper, corn earworm, cotton leafhoppers, cotton leaf perforator, cutworm, green clover worm, lygus, Mexican bean beetle, pink bollworm, podworms, soybean looper, stinkbugs, tobacco budworm, velvetbean caterpillar, woolybean caterpillar, fleabeetle, tomato fruitworm, leafhoppers, loopers.

TRALOMETRIN, SCOUT X-TRA, SAGA

Mfg:Agr Evo

Armyworms, bean beetle, boll weevil, bollworm, corn earworm, cotton leaf perforator, cotton leafworm, cutworms, European cornborer, fleahoppers, grass hoppers, green cloverworms, leafhoppers, loopers, lygus, Mexican bean beetle, pink bollworm, plantbug, stinkbug, thrips, tobacco budworm, velvetbean, velvetbean caterpiller, cockroaches, fleas.

TRICHLORFON, DIPTEREX, DYLOX, PROXOL

Mfg: Miles/Agr Evo

Alfalfa plant bug, alfalfa webworm, armyworm, bagworms, banana caterpillar, beet armyworm, beet webworm, birch leafminer, black fleahopper, budworms, bulb fly, California orange dog, cattle grubs, chinch bug, citrus looper, climbing cutworms, cockroaches, cotton fleahopper, cotton leaf perforator, cotton leafworm, crickets, cutworms, darkling ground beetle, diamondback moth, dipterous leafminer, elm leaf beetle, fleas, forest tent caterpillar, green June beetle, gypsy moth, hornfly, hornworm, horse bots, houseflies, imported cabbageworm, leafhoppers, leafminers, leafrollers, lice, lygus, Mexican bean beetle, Nantucket pine tip moth, Narcissus bulb fly, orange tortrix, pepper maggot, salt marsh caterpillar, seawigs, sod webworms, squash bugs, stinkbug, tarnished plant bug, tent caterpillars, thrips, ticks, tobacco budworm, tomato hornworm, varigated cutworm, webworms, Western bean cutworm, Western yellow striped armyworm, Zimmerman pine tip moth, white grubs, mole crickets.

TRIMETHACARB, BROOT

Mfg: Drexel

Corn rootworm.

ZETA-CYPERMETHRIN, FURY

Mfg: FMC

Cutworms, thrips, boll weevil, cabbage looper, cotton bollworm, leafhoppers, leaf perforator, European corn borer, armworms, pink bollworm, saltmarsh caterpillar, tarnished plant bug, tobacco budworm, plant bugs, lygus, cotton aphid, whiteflies, cucumber beetle, diamondback moth, flea beetles, cabbageworm, leafhoppers, pecan aphids, hickory shuckworm, nut case bearer, pecan weevil.

HERBICIDES

TRADE NAME CONVERSION TABLE — HERBICIDES

Aatrex	*Atrazine*
Abolish	*Thiobencarb*
Accent	*Nicosulfuron*
Accelerate	*Endothall*
Acclaim	*Fenoxaprop-Ethyl*
Accord	*Glyphosate*
Alanap	*Naptalan*
Ally	*Metsulfuron-methyl*
Amber	*Trisulfuron*
Amitrole	*Amino Triazole*
Ansar	*MSMA*
Aquathol	*Endothal*
Arsenal	*Imazapyr*
Assert	*Imazamethabenz-Methyl*
Assure	*Quizalofop-Ethyl*
Assure II	*Quizalofop-P-ethyl*
Asulox	*Asulam*
Avenge	*Difenzoquat*
Balan	*Benefin*
Banvel	*Dicamba*
Barrage	*2,4-D*
Barricade	*Prodiamine*
Basagran	*Bentazon*
Beacon	*Primisulfuron-methyl*
Betanex	*Desmedipham*
Betasan	*Bensulide*
Bladex	*Cyanazine*
Blazer	*Acifluorfen-Sodium*
Bolero	*Thiobencarb*
Brush-Rhap	*2,4-D*
Buctril	*Bromoxynil*
Bueno	*MSMA*
Bugle	*Fenoxaprop-ethyl*
Butoxone	*2,4-DB*
Butyrac	*2,4-DB*
Caparol	*Prometryne*
Casoron	*Dichlorbenil*
Chipton	*Linuron*
Chopper	*Imazapyr*
Clarity	*Dicamba*
Classic	*Chlorimuron-Ethyl*
Cobra	*Lactofen*
Command	*Clomazone*
Cotoran	*Floumeturon*

Cotton-Aide	*Cacodylic acid*
Cropstar	*Alachlor*
Cyclone	*Paraquat*
Dacthal	*DCPA*
Defend	*Bentazon-sodium*
Defol	*Sodium Chlorate*
Demoss	*Fatty Acids*
Devrinol	*Napropamide*
Dimension	*Dithiopyr*
Dr. Biosedge	*Puccinia cannliculata*
Dual	*Metolachlor*
Endurance	*Prodiamine*
Enquick	*Sulfcarbamide*
Entry	*Bentazon*
Eptam	*EPTC*
Eradicane	*EPTC*
Eraser	*Fatty Acids*
Escort	*Metsulfuron-methyl*
Esteron	*2,4-D*
Evik	*Ametryne*
Evital	*Norflurazon*
Express	*Tribenuron-methyl*
Fargo	*Triallate*
Finale	*Glufosinate-ammonium*
Formula 40	*2,4-D*
Frontier	*Dimethenamid*
Fusilade	*Fluaziprop-Butyl*
Gallery	*Isoxaben*
Garlon	*Triclopyr*
Glean	*Chlorsulfuron*
Glyfos	*Glyphosate*
Goal	*Oxyfluorfen*
Gramoxone	*Paraquat*
Grandstand	*Triclopyr*
Graslan	*Tebuthiuron*
Harmony Extra	*Thifensulfuron-Methyl*
Harness	*Acetochlor*
Herbicide 273	*Endothall*
Hoelon	*Diclofop-Methyl*
Honcho	*Glyphosate*
Horizon	*Fenoxaprop-Ethyl*
Hydrothol	*Endothal*
Hyvar	*Bromacil*
Ignite	*Glufosinate-ammonium*
Illoxan	*Diclofop-methyl*
Image	*Imazaquin*

Judge	Alachlor
Jury	Glyphosate
Karmex	Diuron
Kerb	Pronamide
Krenite	Fosamine
Lasso	Alachlor
Lentagran	Pyridate
Lexone	Metribuzin
Linex	Linuron
Londax	Bensulfuron-Methyl
Lontrel	Clopyralid
Lorox	Linuron
Medal	Metolachlor
Meturon	Flometuron
Montar	Cacodylic Acid
Norosac	Dichlorbenil
Nortron	Ethofumesate
Option	Fenoxaprop-Ethyl
Ordram	Molinate
Oust	Sulfometuron-methyl
Pendulum	Pendimethalin
Pennant	Metolachlor
Phytar	Cacodylic Acid
Pinnacle	Thifensulfuron-Methyl
Pledge	Bentazon-sodium
Poast	Sethoxydim
Pramitol	Prometone
Predict	Norflurazon
Prefar	Bensulide
Princep	Simazine
Prism	Clethodim
Prograss	Ethofumesate
Prostar	Propanil
Protocol	Glyphosate
Prowl	Pendimethalin
Pursuit	Imazethapyr
Pyramin	Pryrazon
Ramrod	Propachlor
Ranger	Glyphosate
Rattler	Glyphosate
Reclaim	Clopyralid
Redeem	Triclopyr
Reflex	Fomesafen
Rely	Triclopyr
Remedy	Triclopyr
Reward	Diquat

Rhomine	*MCPA*
Ro-Neet	*Cycloate*
Rodeo	*Glyphosate*
Ronstar	*Oxadiazon*
Roundup	*Glyphosate*
Salvo	*2,4-D*
Savage	*2,4-D*
Scepter	*Imazaquin*
Scythe	*Pelargonic acid*
Select	*Clethedim*
Sencor	*Metribuzin*
Sinbar	*Terbacil*
Solicam	*Norflurazon*
Solution	*2,4-D*
Sonalan	*Ethalfuralin*
Sonar	*Fluridone*
Spike	*Tebuthiuron*
Spin-Aid	*Phenmediphan*
Stam	*Propanil*
Starfire	*Paraquat*
Stinger	*Clopyralid*
Stomp	*Pendimethalin*
Surflan	*Oryzalin*
Surpass. EC	*Acetochlor*
Sutan	*Butylate*
Tackle	*Acifluorfen-Sodium*
Telar	*Chlorsulfuron*
Thistol	*MCPB*
Tillam	*Pebulate*
Tordon	*Picloram*
Torpedo	*Sethoxydim*
Touchdown	*Glyphosate-trimesuim*
Tough	*Pyridate*
Treflan	*Trifluralin*
Tri-4	*Trifluralin*
Trilin	*Trifluralin*
Trust	*Trifluralin*
Tupersan	*Siduron*
Turflon	*Triclopyr*
Vanquish	*Dicamba*
Vantage	*Sethoxydim*
Velpar	*Hexazinone*
Vernam	*Vernolate*
Weedar	*2,4-D*
Weedone	*2,4-D*
Weed-Hoe	*MSMA*
Wham	*Propanil*
Whip	*Fenoxaprop-Ethyl*
Zorial	*Norflurazon*

HERBICIDES

ACEROLA
Paraquat
Roundup
Touchdown

ALFALFA
Bromoxyn
EPTC
Karmex
Kerb
MCPA
Paraquat
Poast
Prowl
Roundup
Sencor
Sinbar
Treflan
2,4-D
Velpar

ALMONDS
Devrinol
EPTC
Fusilade
Goal
MCPA
Paraquat
Poast
Prowl
Roundup
Simazine
Snapshot
Snapshot TG
Solicam
Surflan
Touchdown
Treflan
2,4-D
Velpar

APPLES
Devrinol
Fusilade
Goal
Karmex
Kerb
Ignite
MCPA
Paraquat
Poast
Prowl
Roundup
Simazine
Sinbar
Snapshot
Snapshot TG
Solicam
Surflan
Touchdown
2,4-D

APRICOTS
Devrinol
Fusilade
Goal
MSMA
Paraquat
Poast
Prowl
Roundup
Solicam
Surflan
Touchdown
Treflan
2,4-D

AQUATIC WEEDS and IRRIGATION DITCHES
Acrolein
Banvel
Casoron
Copper Compounds
Copper Sulfate
Curtrine
Diquat
Endothal
Garlon
Karmex
Petroleum Solvents
Rodeo
Roundup
Sonar
Spike
2,4-D
Xylene

ARTICHOKES
Devrinol
Goal
Karmex
Kerb
Poast
Touchdown

ASPARAGUS
Banvel
Devrinol
EPTC
Fusilade
Karmex
Lorox
Paraquat
Poast
Prowl
Roundup
Sencor
Sinbar
Solicam
Touchdown
Treflan
2,4-D

ATEMOYA
Roundup

AVOCADOS
Devrinol
Fusilade
Goal
Paraquat
Poast
Roundup
Simazine
Solicam
Surflan
Touchdown

BANANAS
Ametryne
Goal
Karmex
Paraquat
Roundup
Simazine

BARLEY
Ally
Amber
Assert
Avadex BW
Avenge
Banvel
Bladex
Bromoxynil
Bronate
Buckle
Curtail
Curtail-M
Express
Finesse
Glean
Harmony-Extra
Hoelon
Karmex
MCPA
Paraquat
Roundup
Sencor
Stampede
Stinger

Treflan
Tordon
2,4-D
2,4-DB

BASIL
Devrinol

BEANS — Dry
Basagran
Dacthal
Dual
EPTC
Lasso
Paraquat
Poast
Pursuit
Prowl
Roundup
Sodium chlorate
Sonalan
Treflan

BEANS — Fava
Poast

BEANS — Lima
Basagran
Dacthal
Dual
Lasso
Paraquat
Poast
Prowl
Pursuit
Roundup
Treflan

BEANS — Snap
Basagran
Dacthal
Dual
Paraquat
Poast
Prowl

Pursuit
Roundup
Treflan

BEECHNUT
Goal

BEETS — Table
Pyramin
Ro-neet
Roundup
Spin-Aid

BLUEBERRIES
Devrinol
Karmex
Kerb
Paraquat
Poast
Roundup
Simazine
Sinbar
Solicam
Surflan
Touchdown
2,4-D
Velpar

BREADFRUIT
Roundup

BROCCOLI
Dacthal
Devrinol
Goal
Paraquat
Poast
Prefar
Roundup
Treflan

BRUSSELS SPROUTS
Dacthal
Devrinol
Poast

Prefar
Roundup
Treflan

BUTTERNUT
Goal

CABBAGE
Dacthal
Devrinol
Dual
Goal
Lentagran
Paraquat
Poast
Prefar
Roundup
Tough
Treflan

CANARY GRASS
MCPA

CANEBERRIES
Devrinol
Fusilade
Karmex
Kerb
Paraquat
Petroleum Solvents
Poast
Roundup
Simazine
Sinbar
Solicam
Surflan
Touchdown
2,4-D

CANISTEL
Roundup

CARAMBOLA
Roundup

CARROTS
Fusilade
Lorox
Metribuzin
Paraquat
Prefar
Roundup
Stoddard Solvent
Treflan

CASHEW
Goal

CASSAVA
Paraquat

CASTOR BEANS
EPTC
Treflan

CAULIFLOWER
Dacthal
Devrinol
Goal
Paraquat
Poast
Prefar
Roundup
Treflan

CELERY
Bolero
Caparol
Lorox
Poast
Roundup
Treflan

CELERIC
Roundup

CHERIMOYA
Roundup

CHERRIES
Devrinol
Fusilade

Goal
Kerb
MSMA
Paraquat
Poast
Prowl
Roundup
Simazine
Snapshot
Snapshot TG
Solican
Surflan
Touchdown
Treflan
2,4-D

CHESTNUT
Goal

CHICORY
Roundup
Treflan

CHINESE CABBAGE
Dacthal
Paraquat

CITRUS
Devrinol
EPTC
Fusilade
Goal
Hyvar
Karmex
Krovar
MSMA
Paraquat
Poast
Prowl
Roundup
Simazine
Snapshot
Snapshot TG
Solicam
Surflan
Torpedo

Touchdown
Treflan
2,4-D

CLOVERS
EPTC
Karmec
Kerb
MCPA
Paraquat
Roundup

COCOA
Goal
Paraquat
Roundup

COCONUTS
Roundup

COFFEE
Fusilade
Goal
Roundup

COLLARDS
Dacthal
Paraquat
Poast
Treflan

CORN
Accent
Ametryne
Atrazine
Banvel
Basagran
Beacon
Bicep
Bladex
Broadstrike-Dual
Bromoxynil
Bronco
Buctril
Bullet
Contour

Cycle
Dual
EPTC
Eradicane
Extrazine
Frontier
Goal
Guardsman
Lariat
Lasso
Lorox
Marksman
Paraquat
Petroleum Solvents
Pinnacle
Poast
Prowl
Pursuit
Pursuit-Plus
Ramrod
Resolve
Resource
Roundup
Sencor
Simazat
Simazine
Sodium Chlorate
Stinger
Surpass
Surpass-100
Sutan
Sutazine
Treflan
Tough
2,4-D

COTTON
Assure
Bladex
Banvel
Cacodylic Acid
Cobra
Command
Croak
Dacthal

DEF
Dropp
DSMA
Dual
Endothal
EPTC
Folex
Fusilade
Goal
Harvade
Karmex
MSMA
Paraquat
Poast
Prometryne
Prowl
Ramrod
Roundup
Select
Sodium chlorate
Treflan
WHIP
Zorial

COWPEAS
(Blackeye Beans,
Southern Peas)
Dacthal
Dual
Poast
Pursuit
Sodium Chlorate
Treflan

CRABAPPLE
Goal
Poast
Touchdown

CRANBERRIES
Alanap
Casoron
Devrinol
Poast
Roundup
Simazine

Touchdown
2,4-D
Zorial

CRESS
Dacthal
Treflan

CROWNVETCH
Kerb

CUCUMBERS
Alanap
Curbit
Dacthal
Paraquat
Poast
Prefar
Roundup
Treflan

CURRANTS
Devrinol
Karmex
Paraquat
Roundup
Simazine
Surflan

DATES
Fusilade
Goal
Poast
Roundup
Touchdown

DILL
Caparol

EGGPLANT
Dacthal
Devrinol
Paraquat
Poast
Roundup

ENDIVE
Bolero
Fusilade
Kerb
Roundup
Treflan

FEIJOA
Goal

FIGS
Devrinol
Fusilade
Goal
Paraquat
Poast
Roundup
Surflan
Touchdown

FILBERTS
Devrinol
Fusilade
Goal
Ignite
Paraquat
Roundup
Simazine
Solicam
Surflan
Touchdown

FLAX
Bromoxynil
MCPA
Poast
Sodium chlorate
Treflan

GARBANZO
Dual
Goal
Poast

GARLIC
Bromoxynil
Dacthal

Enquik
Fusilade
Paraquat
Poast
Prowl
Roundup

GENIP
Roundup

GINSENG
Touchdown

GOOSEBERRIES
Karmex
Paraquat
Roundup
Surflan
Touchdown

GRAPES
Devrinol
Fusilade
Goal
Ignite
Karmex
Kerb
MSMA
Paraquat
Poast
Prowl
Roundup
Simazine
Snapshot
Snapshot TG
Solicam
Surflan
Touchdown
Treflan
2,4-D

GRASSES
Ally
Avadex BW
Banvel

Betasan
Bromoxynil
Curtail
Enquik
Garlon
Graslan
Horizon
Karmex
MCPA
MSMA
Nortron
Paraquat
Prograss
Reclaim
Redeem
Remedy
Roundup
Sencor
Stinger
Tordon
2,4-D
Velpar

GUAR
Paraquat
Sodium Chlorate
Treflan

GUAVAS
Atrazine
Fusilade
Goal
Paraquat
Roundup
Surflan
Touchdown

HICKORY NUTS
Goal

HOPS
Endothal
Paraquat
Solicam
Treflan

HORSERADISH
Dacthal
Goal
Roundup

HUCKLEBERRY
Karmex
Paraquat
Roundup
Touchdown

JABOTICABA
Roundup

JACK FRUIT
Roundup

**JERUSALEM
ARTICHOKE**
Roundup

JOJOBA
Goal
Touchdown

KALE
Dacthal
Poast
Roundup
Treflan

KIWI
Devrinol
Goal
Paraquat
Roundup
Surflan
Touchdown

KOHLRABI
Poast

LEEKS
Enquik
Poast

LENTILS
Avadex BW
Far-Go
Paraquat
Poast
Prowl
Pursuit
Ramrod
Roundup
Sencor
Treflan

LESPEDEZA
EPTC
MCPA

LETTUCE
Bolero
Dacthal
Kerb
Paraquat
Poast
Prefar
Roundup

LOQUAT
Goal

LUPINE
Dual
Lorox
Poast
Prowl

**MACADAMIA
NUTS**
Atrazine
Fusilade
Goal
Ignite
Karmex
Paraquat
Poast
Roundup
Simazine

Surflan
Touchdown

MANGOES
Roundup
Touchdown

MARJORAN
Devrinol

MELONS
Alanap
Curbit
Dacthal
Paraquat
Poast
Prefar
Roundup
Treflan

MILLETS
Banvel
2,4-D

MINT
Basagran
Bromoxynil
Enquik
Goal
Karmex
Paraquat
Sinbar
Stinger

MUNG BEANS
Dacthal
Dual
Poast
Treflan

MUSTARD
Dacthal
Poast
Roundup
Treflan

NECTARINES
Devrinol
Fusilade
Goal
Kerb
Paraquat
Poast
Prowl
Roundup
Simazine
Snapshot
Snapshot TG
Solicam
Surflan
Touchdown

NONCROP AREAS
Access
Ametryn
Amino Triazole
Arsenal
Assure
Asulox
Banvel
Barricade
Basagran
Borates
Bromoxynil
Buctril
Cacodylic Acid
Casoron
Crossbow
Curtail
Curtail-M
Demoss
Diquat
Dr. Biosedge
DSMA
Endothal
Endurance
Fatty Acids
Finale
Fusilade
Gallery
Garlon

Goal
Horizon
Hyvar
Ignite
Karmex
Krenite
Krovar
Lontrel
Lorox
MCPA
MCPP
MSMA
Metribuzin
Moncide
Option
Oust
Paraquat
Pathway
Pennant
Poast
Pramitol
Predict
Prometryne
Redeem
Remedy
Rifle
Rodeo
Ronstar
Roundup
Scythe
Simizat
Snapshot
Snapshot TG
Sodium Borates
Sodium Chlorate
Solicam
Spike
Stinger
Stomp
Surflan
Telar
Tordon
Topside
Touchdown
Treflan

2,4-D
2,4-DP
Velpar
Weedmaster
Whip
Zinc Chloride

OATS
Banvel
Bladex
Bromoxynil
Bronate
Curtail
Curtail-M
Glean
Harmony-Extra
Lorox
MCPA
Paraquat
Propanil
Roundup
Stinger
Tordon
2,4-D

OKRA
Roundup
Treflan

OLIVES
Devrinol
Goal
Karmex
Paraquat
Poast
Roundup
Simazine
Snapshot
Snapshot TG
Surflan
Touchdown

ONIONS
Brominal
Dacthal

Enquik
Fusilade
Goal
Paraquat
Poast
Prowl
Prefar
Roundup
Sulfuric Acid
Treflan

ORNAMENTALS
Acclaim
Alanap
Amino Triazole
Asulox
Barricade
Bensulfide
Casoron
Dacthal
Derby
Devrinol
Endurance
EPTC
Fatty Acids
Fusilade
Gallery
Goal
Ignite
Karmex
Kerb
Lasso
Lorox
Ornamental Herbicide II
Paraquat
Pennant
Petroleum Solvents
Poast
Predict
Pro-Grow
Redeem
Ronstar
Roundup
Rout
Scythe

Simazat
Simazine
Snapshot
Snapshot TG
Stinger
Stomp
Surflan
Touchdown
Treflan
Vantage
Vapam
XL

PAPAYA
Goal
Karmex
Paraquat
Roundup
Surflan
Touchdown

PARSLEY
Lorox

PARSNIPS
Lorox
Roundup

PASSION FRUIT
Paraquat
Touchdown

PEACHES
Devrinol
Fusilade
Goal
Karmex
Kerb
MSMA
Paraquat
Poast
Prowl
Roundup
Simazine
Sinbar

Snapshot
Snapshot TG
Solicam
Surflan
Touchdown
Treflan
2,4-D
Zorial

PEANUTS
Balan
Basagran
Blazer
Classic
Dual
Enquik
Lasso
Paraquat
Poast
Prowl
Pursuit
Roundup
Sonalan
Tackle-Plus
Tough
Treflan
Vernam
Whip
2,4-DB
Zorial

PEARS
Devrinol
Fusilade
Goal
Karmex
Kerb
MSMA
Paraquat
Poast
Prowl
Ronstar
Roundup
Simazine
Sinbar

Snapshot
Snapshot TG
Solicam
Surflan
Touchdown
2,4-D

PEAS
Avadex BW
Basagran
Buckle
Command
Dual
Karmex
MCPA
MCPB
Paraquat
Poast
Pursui
Roundup
Sencor
Sonalan
Treflan

PECANS
Devrinol
Fusilade
Goal
Ignite
Karmex
Paraquat
Poast
Ronstar
Roundup
Simazine
Sinbar
Snapshot
Snapshot TG
Solicam
Surflan
Touchdown
Treflan

PEPPERS
Basagran
Command
Dacthal
Devrinol
Dual
Fusilade
Paraquat
Poast
Prefar
Roundup
Sodium chlorate
Treflan

PERSIMMONS
Devrinol
Goal
Roundup
Touchdown

PIGEON PEA
Paraquat
Prometryn

PINEAPPLE
Ametryne
EPTC
Hyvar
Karmex
Paraquat
Roundup
Simazine
Velpar

PISTACHIO
Devrinol
Fusilade
Goal
Ignite
Paraquat
Poast
Prowl
Ronstar
Roundup
Surflan

77

Touchdown
2,4-D

PLUMS
Devrinol
Fusilade
Goal
Kerb
MSMA
Paraquat
Poast
Prowl
Roundup
Simazine
Snapshot
Snapshot TG
Solicam
Surflan
Touchdown
Treflan
2,4-D

POMEGRANATES
Devrinol
Fusilade
Goal
Poast
Roundup
Snapshot
Snapshot TG
Surflan
Touchdown

POTATOES
Dacthal
Diquat
Dual
Endothal
Enquik
EPTC
Karmex
Lorox
Paraquat
Poast
Prowl

Roundup
Sencor
Sulfuric Acid
Treflan
Turbo
2,4-D

PRUNES
Devrinol
Fusilade
Goal
Kerb
MSMA
Paraquat
Poast
Prowl
Roundup
Simazine
Solicam
Surflan
Touchdown
Treflan
2,4-D

PUMPKINS
Command
Paraquat
Poast
Prefar
Roundup

QUINCE
Goal
Poast
2,4-D

RADISHES
Dacthal
Roundup
Treflan

**RANGELAND and
PASTURES**
Ally
Banvel

Crossbow
Curtail
Graslan
Grazon
Paraquat
Reclaim
Redeem
Remedy
Roundup
Stinger
Tordon
2,4-D
2,4-DP
Velpar
Weedmaster

RAPE
Poast
Treflan

RHUBARB
Devrinol
Fusilade
Kerb
Paraquat
Poast

RICE
Arrosolo
Basagran
Blazer
Bolero
Copper Derivatives
Copper Sulfate
Facet
Hydrothal
Londax
MCPA
Ordram
Propanil
Prowl
Roundup
Sodium Chlorate
2,4-D
Triclopyr
Whip

ROSEMARY
Devrinol

RUTABEGAS
Dacthal
Roundup

RYE
Bromoxynil
Bronate
Karmex
Lorox
MCPA
Paraquat
2,4-D

SAFFLOWER
Avadex
Dual
EPTC
Paraquat
Sodium Chlorate
Treflan

SALISFY
Roundup

SAINFOIN
Kerb
Poast
Sencor
Sinbar

SAPOTE
Roundup

SAVORY
Devrinol

SORGHUM (Milo)
Atrazine
Banvel
Basagran
Bicep

Bromoxynil
Bronco
Buctril
Bullet
Cycle
Dual
Harness
Karmex
Laddok
Lasso
Lorox
Marksman
MCPA
Paraquat
Prowl
Ramrod
Roundup
Sodium Chlorate
Surpass
Treflan
2,4-D

SOURSOP
Roundup

SOYBEANS
Assure
Basagran
Bayonet
Blazer
Broadstrike-Dual
Broadstrike-Treflan
Bronco
Canopy
Classic
Cobra
Collego
Command
Commence
Concert
Detail
Dual
Freedom
Frontier
Fusilade

Fusion
Galaxy
Gemni
Goal
Harmony
Harness
Lasso
Lexone
Lorox
Lorox-Plus
MCPA
Option
Paraquat
Passport
Pinnacle
Poast
Preview
Prowl
Pursuit
Pursuit-Plus
Ramrod
Reflex
Resource
Roundup
Salute
Scepter
Scepter-O.T.
Select
Sencor
Sodium chlorate
Sonalan
Squadron
Storm
Surpass
Synchrony-STS
Tornado
Treflan
Tri-Scept
Turbo
Typhoon
Vernam
2,4-D
2,4-DB
Zorial

SPICES
Devrinol

SPINACH
Poast
Ro-Neet
Roundup
Spin-Aid

SQUASH
Command
Dacthal
Paraquat
Poast
Prefar
Roundup
Treflan

STRAWBERRIES
Dacthal
Devrinol
EPTC
Paraquat
Poast
Roundup
Simazine
2,4-D

SUGAR APPLE
Roundup

SUGAR BEETS
Betamix
Betanex
Endothal
EPTC
Lontrel
Nortron
Paraquat
Poast
Pyramin
Ramrod
Ro-Neet
Roundup
Stinger

Tillam
Treflan

SUGARCANE
Ally
Ametryne
Asulam
Atrazine
Banvel
Diquat
Karmex
Paraquat
Prowl
Roundup
Sencor
Sinbar
Treflan
2,4-D
Velpar
Weedmaster

SUNFLOWERS
Assert
EPTC
Paraquat
Poast
Prowl
Sodium Chlorate
Sonalan
Treflan

SWEET POTATOES
Command
Dacthal
Devrinol
Fusilade
Paraquat
Poast
Roundup

TAMARIND
Roundup

TANIER
Paraquat

TARO
Goal
Paraquat

TOBACCO
Balan
Command
Devrinol
Edge
Poast
Prowl
Tillam

TOMATOES
Dacthal
Devrinol
Eptam
Metribuzin
Paraquat
Poast
Roundup
Tillam
Treflan

TREFOIL
Balan
EPTC
Karmex
Kerb
MCPA
Paraquat
Roundup
2,4-DB

TRITICALE
Bromoxynil

TURF
Acclaim
Amino Triazole
Asulox
Atrazine
Balan
Banvel
Barricade
Basagran

Bensulfide
Buctril
Cacodylic Acid
Calar
Confront
Dacthal
Devrinol
Dimension
DMC
DSMA
Fatty Acids
Ferric Sulfate
Gallery
Garlon
Illoxan
Image
Kerb
Lexone
MCPA
MCPP
MSMA
Nortron
Pennant
Prograss
Prompt
Prowl
Ronstar
Roundup
Sencor
Simazine
Slythe
Surflan
Team
Tupersan
Turflon
Turflon II
2,4-D
2,4-DP
Vapam
Vorlex
X-L

TURNIPS
Dacthal
Devinol

Paraquat
Poast
Roundup
Treflan

TYFON
Paraquat

VETCH
Karmex
MCPA

WALNUTS
Devrinol
EPTC
Fusilade
Goal
Ignite
Karmex
MSMA
Paraquat
Poast
Prowl
Roundup
Simazine
Solicam
Surflan
Touchdown
Treflan

WATERCRESS
Roundup

WATERMELONS
Alanap
Curbit
Dacthal
Paraquat
Poast
Prefar
Roundup
Treflan

WHEAT
Ally
Amber
Assert
Atrazine
Avadex BW
Avenge
Banvel
Bronate
Bladex
Bromoxynil
Buckle
Cheyenne
Command
Curtail
Curtail-M
Dakota
Express
Fallowmaster
Finesse
Glean
Harmony
Harmony-Extra
Hoelon
Karmex
Landmaster BW
Landmaster II
Lorox
MCPA
Paraquat
Propanil
Prowl
Roundup
Sencor
Stampede
Stinger
Surflan
Tiller
Tordon
Tornado
Treflan
2,4-D

HERBICIDES

2,4-D   **Mfg: Numerous**

Alder, alkali mallow, alligatorweed, arrowhead, artesmesia, artichoke, ashers, Austrian fieldcress, beggarsticks, big sagebrush, bindweed, bitter sneezeweed, bitter watercress, bitterweed, blackeye susan, bladderwort, blessed thistle, blue lettuce, blue weed, brambles, broomweed, buckbrush, bull thistle, buckrush, burdock, bur ragweed, Canada thistle, carelessweed, carpetweed, catnip, Cherokee rose, chickweed, chicory, coastal sagebrush, cocklebur, coffee weed, coontail, corncockle, corn spurry, croton, curly dock, curly indigo, dandelion, dayflower, dock, dogbane, duck salad, elderberry, elodea, evening primrose, fanweed, fanwort, fiddleneck, fleabane, flower-of-an-hour, french weed, galinsoga, goatsbeard, goldenrod, gooseweed, ground ivy, gumweed, hazel heal-all, hemp, henbit, hoary cress, horse nettle, horsetail, Japanese honeysuckle, Johnson weed, ironweed, knotweed, lambsquarters, leafy spurge, liveoak, locoweed, madrone, mallow, manzanita, marestail, marshelder, Mexican weed, milkweed, morningglory, mustard, nettles, oak, orange hawkey, pennycress, pennywort, peppergrass, pepperwood, pigweed plantain, poison hemlock, poor joe, povertyweed, prickly lettuce, primrose, puncturevine, rabbit brush, ragweed, redstem, Russian knapweed, sages, sand sagebrush, St. John's wort, shepherd's-purse, slim asters, smartweed, snow-on-the-mountain, snow-on-the-prairie, sowthistle, spikerush, spurge, stinging nettle, sumac, sunflower, tall indigo, tanoak, Tansy ragwort, tan weed, tarweed, tea vine, Texas blueweed, toad flax, turkey pea, umbrella sedge, velvetleaf, vervains, vetch, Virginia creeper, water milfoil, water naiad, wild carrot, winter cress, wild cucumber, wild garlic, wild grapes, wild lettuce, wild onion, wild parsnip, wild radish, willow, Yankee weed, yellow rocket, yellow star thistle.

2,4-DP, DICHORPROP   **Mfg: Rhone Poulenc**

Oaks, pine, fir, spruce, black cherry, alder, willow, sansage, elm and most annual and perennial broadleaf weeds.

2,4-DB, **BUTYRAC, BUTOXONE**   **Mfg: Rhone Poulenc**
                                                                      and Cedar Chem.

Bull thistle, Canada thistle, cocklebur, coffeeweed, curly dock, deadnettle, devil's claw, fanweed, fiddleneck, filaree, goatweed, groundsel, hairy catsear, hedge bindweed, jimsonweed, knotweed, kochia, lambsquarters, London rocket, morningglory, mustards, nightshade, pigweed, plantain, prickly lettuce, ragweed, Russian thistle, Shepherd's-purse, smartweed, stinkweed, sugar beets, sweet clover, tansy mustard, teaweed, velvetleaf, Virginia coffeeleaf, white top, wild beet, wild turnip, yellow rocket, penny cress, sicklepod.

ACCESS (picloram/triclopyr)　　　　　　　Mfg: DowElanco

Ash, aspen, buck cherry, elm, hackberry, hickory, locust, maple, multiflora rose, oceanspray, pine, poplar, sassafras, tan oak.

ACETOCHLOR, SURPASS EC,　　　　Mfg: Monsanto/Zeneca
HARNESS

Nightshade, carpetweed, ragweed, beggarweed, Florida pusley, galensoga, kochia, lambsquarters, pigweed, prickly sida, purslane, smartweed, tall waterbugs, barnyardgrass, bristly foxtail, broadleaf signalgrass, panicum, crabgrass, crow footgrass, sandbur, foxtail millet, foxtails, goosegrass, cupgrass, red rice, sprangletop, johnsongrass, shattercase, wild prose millet, witchgrass, yellow nutsedge.

ACIFLUORFEN-SODIUM,　　　　Mfg: Rhone Poulenc and BASF
BLAZER, TACKLE

Amaranth, balloonvine, bindweed, bristly starbur, buffalobur, burghekin, Canada thistle, carpetweed, citron, cocklebur, coffeeweed, coffee senna, copperleaf, croton, devil's claw, fall panicum, Florida pusley, foxtails, galinsoga, giant ragweed, groundcherry, hairy indigo, hemp sesbania, hop hornbeam, jimsonweed, lady's thumb, lambsquarters, milkweed, morningglory, mustard, nightshade, pigweed, poorjoe, prostrate spurge, purslane, ragweed, redvine, seedling johnsongrass, shattercane, tall waterhemp, showy crotalaria, smell melon, smartweed, smooth pigweed, Texas gourd, trumpet creeper, velvetleaf, Virginia copperleaf, volunteer cowpea, volunteer small grains, wild buckwheat, wild cucumber, wild mustard, wild poinsettia.

ALACHLOR, LASSO,　　　　　　　　Mfg: Monsanto
JUDGE, CROPSTAR

Barnyard grass, black nightshade, carpetweed, crabgrass, fall panicum, Florida pusley, foxtails, goosegrass, hairy nightshade, pigweed, purslane, red rice, seedling signal grass, witchgrass, yellow nutgrass.

AMETRYNE, EVIK　　　　　　　　　Mfg: CIBA

Annual broadleaves, annual grasses, Bachiaria, barnyardgrass, cocklebur, crabgrass, crotalaria, dallisgrass, fireweed, Flora's paintbrush, Florida pusley, foxtails, goosegrass, Japanese tea, jungle rice, kukaipuaa, lambsquarters, morningglory, mustard, nutsedge, panicum, pigweed, purslane, ragweed, rattlepod, shattercane, signalgrass, sowthistle, Spanish needles, velvetleaf, wild pea bean, wild proso millet, wiregrass.

AMINO TRIAZOLE, AMITROLE   Mfg: Rhone Poulenc

Annual bluegrass, ash, barnyardgrass, bermudagrass, big leaf maple, blackberry, bluegrasses, Canada thistle, carpetweed, catchfly, cattails, cheatgrass, chickweed, chrysanthemum weed, cocklebur, crabgrass, dandelion, dock, downybrome, fanweed, foxtail, goosegrass, groundsel, hempnettle, honeysuckle, horsenettle, horsetail rush, jimsonweed, knotweed, kochia, kudzuvine, lambsquarters, leafy spurge, locust milkweed, mustard, nightshade, nutgrass, phragmites, pigweed, plaintain, poison ivy, poison oak, puncturevine, purslane, quackgrass, ragweed ryegrass, salmon berry, Shepherdspurse, smartweed, sowthistle, stinkweed, sumac, sunflower, tarweed, velvetgrass, volunteer alfalfa, water hyacinth, white cockle, white grass, whitetop, wild barley, wild buckwheat, wild cherry, wild oats, yellow rocket.

ARROSOLO (molinate/propanil)   Mfg: Zeneca

Barnyard grass, broadleaf signalgrass, cockspur, crabgrass, curly dock, day flower, hemp sesbania, hoorahgrass, jingle rice, kornel beakrush, paragrass, pigweed, redweed, spikerush, sprangletop, Texas millet, Texasweed, wooly croton.

ASULAM, ASULOX   Mfg: Rhone Poulenc

Alexandergrass, barnyardgrass, broadleaf panicum, bullgrass, California grass, crabgrass, foxtails, goosegrass, guineagrass, horseweed, itchgrass, Johnsongrass, panicum, paragrass, Raoulgrass, sandbur, Western bracken.

ATRAZINE, AATREX   Mfg: CIBA and others

Ageratum, amaranth, barnyardgrass, blackeyed susans, bluegrass, broomsedge, broomweed, burdock, Canada thistle, carpetweed, cheatgrass, chickweed, cinquefoil, cocklebur, crabgrass, cranesbell, dock, dogfennel, dogbane, downybrome, fanweed, fiddleneck, fireweed, fleabane, Flora's paintbrush, foxtail, galinsoga, goosegrass, groundcherry, groundsel, hoary cress, horseweed, jimsonweed, jungle rice, knotweed, kochia, lady's thumb, lambsquarters, little barley, little bluestern, medusahead, morningglory, mustard, nightshade, nutgrass, orchardgrass, papalo, pellitory weed, pigweed plantain, pineapple weed, poverty weed, puncturevine, purple top, purslane, quackgrass, ragweed, rattlepod, redtop, Russian thistle, sagewort, sandbur, shepherd's-purse, sicklepod, smartweed, smooth brome, sneezeweed, sow thistle, Spanish needles, spurge, sunflower, tall buttercup, tansy mustard, tumble mustard, velvetleaf, volunteer grains, watergrass, wild carrot, wild lettuce, wild oats, wiregrass, wirestem, witchgrass, yarrow.

BENEFIN, BALAN   Mfg: Platte Chemical Co.

Annual bluegrass, barnyardgrass, carelessweed, carpetweed, chickweed, crabgrass, crowfootgrass, deadnettle, Florida pusley, foxtail, goosegrass, Johnson grass (from seed), jungle rice, knotweed, lambsquarters, pigweed, purslane, redmaids, ryegrass, sandbur, shepherd's purse, Texas panicum.

BENSULFURON-METHYL,   Mfg: DuPont
LONDAX

Annual arrowweed, blunt spikerush, dayflower, ducksalad, ecilpta, Eisen waterhyssop, false pimernel, gooseweed, Mexicanweed, mud poaintan, pickerelweed, purple ammania, redstem, rice flatsedge, roughseed bubrush, Southern naiad, Texasweed, waterplaintain, waterwort, yellow nutsedge, roundleaf waterhyssop, roundleaf, waterhyssop.

BENSULIDE, BETASAN,   Mfg: Zeneca/Gowan
PREFAR

Annual bluegrass, barnyardgrass, crabgrass, deadnettle, fall panicum, foxtails, goosegrass, jungle rice, lambsquarters, pigweed, purslane, shepherd's-purse, springletop, watergrass.

BENTAZON, BASAGRAN,   Mfg: BASF
PLEDGE, DEFEND, ENTRY

Annual sedges, arrowhead, balloon vine, beggarticks, bristly starbur, Canada thistle, cocklebur, coffee senna, day flower, devils claw, duck salad, galinsoga, giant ragweed, hairy nightshade, jimsonweed, lady's thumb, lambsquarters, morningglory, pigweed, prickly sida, pruple nutsedge, purslane, ragweed, redstem, redweed, river bulrush, sesbania, shepherd's-purse, smartweed, spike rush, spurred anoda, sunflower, tropic croton, velvetleaf, venice mallow, water plantain, wild buckwheat, wild mustard, wild poinsettia, yellow nutgrass.

BETAMIX (phenmediphan/ desmediphan)

Mfg:Agr Evo

Annual sowthistle, chickweed, fiddleneck, foxtails, goosefoot, groundcherry, kochia, lambsquarters, London rocket, nightshade, pigweed, purslane, ragweed, shepherd's purse, wild buckwheat, wild mustard.

BICEP (atrazine/metolachlor)

Mfg: CIBA

Barnyardgrass, black nightshade, carpetweed, Chickweed, henbit, cocklebur, crabgrass, crowfootgrass, cupgrass, Florida pusley, foxtails, galinsoga, goosegrass, hairy nightshade, jimsonweed, kochia, lambsquarters, morningglory, mustard, panicum, pigweed, prickly sida, pruslane, ragweed, red rice, signalgrass, smartweed, velvetleaf, witchgrass, yellow nutsedge.

BORATES

Mfg: U.S. Borax and others

Bermudagrass, bindweed, Canada thistle, dogbane, Johnson grass, Klamath weed, leafy spurge, poison ivy, poison oak, toad flax, wild lettuce, white top, and annual weeds and grasses.

BROADSTRIKE + DUAL (flumetsulan/ metolachlor)   Mfg: DowElanco

Barnyardgrass, crabgrass, crowfootgrass, cupgrass, fall panicum, foxtails, goosegrass, johnsongrass, wild proso millet, yellow nutsedge, red rice, sandbur, shattercane, signalgrass, volunteer sorghum, witchgrass, spurrel anoda, palmer amaranth, beggarweed, cocklebur, chickweed, carpetweed, goosefoot, henbit, kochia, lambsquarters, ladysthumb, Venice mallow,. mustard, morninglory, nightshade, pigweed, puncturevine, purslane, Florida pusley, ragweed, Russian thistle, sicklepod, prickly sida, smartweed, spurge, sunflower, velvetleaf, tall waterhemp.

BROADSTRIKE + TREFLAN (flumetsulan/trifluralin)

Mfg: DowElanco

Annual bluegrass, barnyardgrass, brachiaria, bromegrasses, cheat, crabgrass, foxtails, guineagrass, johnsongrass, jungle rice, wild oats, fall panicum, ryegrass, Texas panicum, sandbur, sprangle top, stinkgrass, shattercane, cupgrass, spurrel anoda, Palmer amaranth, beggarweed, carpetweed, chickweed, cocklebur, goosefoot, henbit, jimsonweed, knotweed, kochia, Venice mallow, morninglory, mustards, nightshade, ladythumb, lambsquarters, pigweed, puncturevine, purslane, Florida pusley, ragweed, Russian thistle, sicklepod, prickly sida, smartweed, spurge, stinging nettle, sunflower, velvetleaf, tall waterhemp.

BROMACIL, HYVAR   Mfg: DuPont

Aster, bahiagrass, balsam apple, barnyard grass, bermudagrass, bluegrass, bouncing bet, bracken fern, bromegrass, bromegrass, broomsedge, bur buttercup, cheatgrass, China lettuce, Coloradograss, cottonweed, cottonwood, crabgrass, crowfoot, dallisgrass, dandelion, dogbane, dog fennel, elms, Flora's paintbrush, Florida pusley, foxtails, goldenrod, goosegrass, hackberry, henbit, Hialoa, horsetail, Johnson grass, jungle rice, lambsquarters, maples, mustard, nutgrass, oaks, orchardgrass, pangolagrass, paragrass, pigweed, pine, plantain, poplar, prostate knotweed, puncturevine, purple top, purslane, quackgrass, ragweed, redbud, redtop, ryegrass, saltgrass, sandbur, sedge, sprangletop, sweetgum, sumac, torpedograss, turkey mullein, wild carrot, wild cherry, wild oats, willow wiregrass, willows, vasey grass.

BROMOXYNIL, BUCTRIL   Mfg: Rhone Poulenc

Annual morningglory, annual nightshade, bachelors button, bassia, black mustard, burchervil, catch weed, corn cockle, cocklebur, corn chomomile, cow cockle, dog fennel, false flax, fanweed, fiddleneck, field pennycress, flixweed, fumitory, green smartweed, giant ragweed, gromwell, groundsel, henbit, Jacobs ladder, jimsonweed, knawel, kochia, lambsquarters, London rocket, mayors tail, miners lettuce, morningglory, peppergrass, pigweed, prickly lettuce, prostrate spurge, puncturevine, purple mustard, ragweed, redmaids, Russian thistle, saltbrush, marestail, shepherd's purse, silverleaf nightshade, smartweed, sunflower, tall waterhemp, tansy mustard, tartary buckwheat, tarweed, tumble mustard, velvetleaf, wild buckwheat, wild mustard, wild radish, winter vetch, yellow rocket.

BRONATE (bromoxynil/MCPA)   Mfg: Rhone Poulenc

Sowthistle, mustard, nightshade, cocklebur, lambsquarters, tarweed, cow cockle, fiddleneck, full pennycress, horned poppy, jimsonweed, ladysthumb, marshelder, London rocket, smartweed, pepperweed, pigweed, shepards purse, sunflower, tall waaterhops, buckwheat, wild buckwheat, yellow rocket, groundsel, ragweed, corn chamomile, corn gromwell, fumitory, sesbania, giant ragweed, henbit, knarvel, kochia, mayweed, knotweed, puncturevine, tansy mustard, velvetleaf, wild radish.

BRONCO (alachlor/glyphosate)   Mfg: Monsanto

Annual bluegrass, downy brone, volunteer corn, crabgrass, foxtails, fall panicum, field sandbur, shattercane, volunteer wheat, fleabane, kochia, lambsquarters, prickly lettuce, marestail, pigweed, ragweed, giant ragweed, smartweed, sunflowers, Russian thistle, velvetleaf, Kentucky bluegrass, smooth brome, fescue, wirestem mukly, johnsongrass, orchardgrass, quackgrass, perennial ryegrass, timothy alfalfa, clovers, curly dock, milkweed, mulleir, Canada thistle, swamp smartweed, barnyardgrass, crabgrass, goosegrass, broadleaf signalgrass, witchgrass, carpetgrass, nightshade, purslane, Florida pusley.

BUCKLE (triallate/trifluralin)   Mfg: Monsanto

Foxtails, downy brome, wild oats.

BUCTRIL + ATRAZINE   Mfg: Rhone Poulenc

Buffalobur, nightshade, bur cucumber, cocklebur, lambsquarter, ragweed, morningglory, giant ragweed, sesbania, jimsonweed, kochia, ladysthumb, smartweed, prickly sida, puncturevine, pigweed, sunflower, tall waterhemp, spurge, velvetleaf, Venice mallow, wild buckwheat, mustard.

BULLET (alachlor/atrazine)   Mfg: Monsanto

Barnyardgrass, broadleaf signalgrass, carpetweed, cocklebur, crabgrass, Florida beggarweed, Florida pusley, foxtails, goosegrass, ground cherry, horseweed, jimsonweed, johnsongrass, kochia, lambsquarters, morningglory, mustard, nightshade, panicum, pigweed, purslane ragweed, red rice, sandbur, shattercane, sicklepod, smartweed, sprangletop, teaweed, volunteer sorghum, witchgrass, yellow nutsedge.

BUTYLATE, SUTAN +   Mfg: Zeneca

Barnyardgrass, bermudagrass seedlings, crabgrass, fall panicum, foxtails, goosegrass, Johnson grass (seedling), purple nutgrass, sandbur, Texas panicum, volunteer sorghum, watergrass, wild cane, yellow nutgrass.

CACODYLIC ACID,   Mfg: Monterey Chemical Co.,
PHYTAR, MONTAR, COTTON AIDE   Pamol

Annual weeds, cotton defoliant.

CANOPY (chlorimuron-ethyl/   Mfg: DuPont
metribuzin)

Cocklebur, common ragweed, copperleaf, Florida beggarweed, giant ragweed, hemp sesbania, hophorn bean, jimsonweed, lambsquarters, morningglory, pigweed, prickly sida, purslane, sicklepod, smartweed, spotted spurge, sunflower, teaweed, velvetleaf.

CHEYENNE (fenoxaprop-ethyl/ MCPA/sulfonyl urea)　　　　Mfg: Agr Evo

Foxtails, foxtail millet, proso millet, wild oats, annual sowthistle, mustards, Canada thistle, chickweed, groundsel, lambsquarters, curly dock, false chamomile, field pennycress, flex weed, smartweed, kochia, ladythumb, London rocket, marshelder, miners lettuce, knotweed, pigweed, Russian thistle, mayweed, buttercups, dogfennel, fiddleneck, volunteer peas and lentils, sunflower, wild buckwheat, wild chamomile, wild garlic.

CHLORIMURON-ETHYL, CLASSIC　　　　Mfg: DuPont

Beggartick, beggarweed, bristly starbur, cocklebur, giant ragweed, Florida beggarweed, hemp sesbania, Jerusalem artichoke, jimsonweed, morningglory, mustard, pigweed, ragweed, sicklepod, smartweed, sunflower, velvetleaf, wild poinsettia, yellow nutsedge.

CHLORSULFURON, GLEAN, TELAR　　　　Mfg: DuPont

American strawberry, annual bluegrass, aster, Australian saltbush, bedstraw, bouncing bet, buckhorn plantain, bull thistle, burbeak-chevil,bur beakchervil, bur clover, buttercup, Canadian thistle, chickweed, cinquefoil, common tansy, conial catchfly, corn spurry, curly dock, dandelion, dog fennel, Dyer's wort, false chamomile, false flax, fiddleneck, field pennycress, filaree, flixweed, foxtails, goldenrod, gromwell, groundsel, henbit, hoarycress, horsetails, knotweed, lady's thumb, lambsquarters, London Rocket, mallow, milk thistle, miners lettuce, mullen, musk thistle, mustards, nettle, pepperweed, pigweed, pineapple weed, poison hemlock, prickly lettuce, puncture vine, ragweed, red clover, Russian kanpweed, ryegrass, scotch thistle, scouring rush, Shepherd's purse, smartweed, sowthistle, speedwell, star thistle, sunflower, sweet clover, tansy mustard, tumble mustard, turley mullen, vetch, white clover, white cockle, white top, wild buckwheat, wild carrot, wild garlic, wild onion, wild parsnips, wild radish, wild turnip, yarrow, yellow starthistle.

CLETHODIM, SELECT, PRISM　　　　Mfg: Valent

Johnsongrass, seltercane, volunteer corn, wild oats, wild proso millet, witchgrass, wooly cupgrass, barnyardgrass, broadleaf signalgrass, crabgrass, crowfoot grass, fall panicum, sandbur, foxtails, goosegrass, itchgrass, jungle rice, sprangletop, red rice, cupgrass, volunteer cereals, bermudagrass, quackgrass, shattercane, Texas panicum, wirestem muhley.

CLOMAZONE, COMMAND   Mfg: FMC

Barnyardgrass, bermudagrass, broadleaf signalgrass, cocklebur, crabgrass, downy brome, Florida beggerweed, Florida pusley, foxtails, goosegrass, jimsonweed, Johnsongrass, jointed goatgrass, kochia, lambsquarters, panicums, pitted morningglory, prickly lettuce, prickly sida, purslane, ragweed, sandbur, shattercane, smartweed, southwestern cupgrass, spurge, spurred anoda, taney mustard, tropic croton, tumble mustard, velvetleaf, venice mallow, volunteer rye, volunteer wheat, wild buckwheat, wild proso millet, wooly cupgrass.

CLOPYRALID, RECLAIM,   Mfg: DowElanco
LONTREL, STINGER

Acacias, annual ragweed, Canadian thistle, chamomile, cocklebur, giant ragweed, knapweed, lady's thumb, mesquite, musk thistle, smartweed, sowthistle, sunflower, vetch, volunteer alfalfa, volunteer soybeans, wild buckwheat, jimsonweed, Jerusalem artichoke, clover, vetch, marsh elder, nightshade, buffalobur, burdock, chamomile, maryweed, coffeeweed, cornflower, dandelion, hawksbeard, horseweed, prickly lettuce, locoweed, oxeye daisy, pineappleweed, salsify, sicklepod, red sorrel, yellow starthistle, volunteer peas, volunteer lentils, volunteer beans.

COMMENCE (clomazone/   Mfg: FMC, DowElanco
trifluralin)

Annual bluegrass, barnyardgrass, Brachiaria, bromegrass, carpetgrass, cheatgrass, chickweed, crabgrass, Florida pusley, foxtails, goosefoot, goosegrass, Johnson grass, jungle rice, knotweed, kochia, lambsquarters, panicum, pigweed, purslane, Russian thistle, sandbur, shattercane, sprangletop, stinging nettle, stinkgrass, velvetleaf, Venice mallow, wooly cupgrass.

CONCERT (trifensulfuron-methyl/   Mfg: DuPont
chlorimuron-ethyl)

Smartweed, cocklebur, jimsonweed, lambsquarters, pigweed, velvetleaf, mustard, sunflowers.

CONFRONT (triclopyr/clopyralid)　　　Mfg: DowElanco

Beggarweed, black media, broadleaf plaintain, buckhorn plaintain, burdock, Canada thistle, cocklebur, coffeeweed, curly dock, dandelion, false dandelion, germander speedwell, goldenrod, ground ivy, henbit, hop clover, lambsquarters, lespedoza, matchweed, musk thistle, poison ivy, ragweed, red clover, shepards purse, smartweed, spotted catsear, spurweed, vetch, Virginia pepperweed, white clover, wild buckwheat.

CONTOUR (imazethapyr/atrazine)　　　Mfg: American Cyanamid

Spurrel anoda, Jerusalem artichoke, buffalobur, bristly starbur, carpetweed, cocklbur, ecilpta, galensoga, jimsonweed, kochia, lambsquarter, Venice mallow, marshelder, morninglory, mustard, nightshade, pigweed, wild poinsettia, puncturvine, purslane, Florida pusley, prickly sida, ragweed, barnyard sage, smartweed, spurge, sunflower, velvetleaf, Canada thistle, Russian thistle, tall waterhemp, barnyard grass, crabgrass, wooly cupgrass, foxtails, goosegrass, johnsongrass, wild proso millet, panicums, red rice, shattercane, broadleaf signalgrass, sorghum, nutsedge, crowfootgrass, itchgrass, sandbur, witchgrass.

COPPER SULFATE AND ITS DERIVATIVES
Mfg: Applied Biochemist, Agtrol, Griffin, Phelps-Dodge, Sandoz & others

Algae, hydrilla and other aquatic weeds.

CROAK (fluometuron/MSMA)　　　Mfg: Drexel

Barnyardgrass, brochiaria, crabgrass, crowfootgrass, dallisgrass, panicum, foxtails, goosegrass, johnsongrass, nutgrass, ryegrass, sandbur, buttonweed, cocklebur, Florida pusley, goathead, jimsonweed, lambsquarters, morningglory, pigweed, prickly sida, puncturevine, purslane, ragweed, sesbania, sicklepod, smartweed, tumbleweed.

CROSSBOW (triclopyr/2,4-D)　　　Mfg: DowElanco

Alder, ash, aspen, bedstraw, beech, black locust, boneset, cascaro, brich, blackberry, buckbrush, bull thistle, bur clover, burdock, buttercup, Canadian thistle, cherry, chickweed, cocklebur, dandelion, docks, elderberry, field pennycress, goldenrod, ground ivy, hazel, honeysuckle, knapweed, lambsquarters, maples, marshelder, milkweed, multiflora rose, musk thistle, oaks, oxalis, pine, plantain, poison oak, pokeweed, ragweeds, sumac, sunflower, sweet gum, tall ironweed, tamarak, vetch, wild buckwheat, wild carrot, willow,beech, boneset, cascara, ceanothus, cottonwood, dogwood, elderberry, hawthorn, poison ivy, sassafras, scotch broom, sumac, sycamore, tanoak, wax myrtle, wild grape, white oak, buck brush, elm, Russian olive, sweetgum bluewood, horseweed, mustard, spurge, bedstraw, bluebur, clovers, woolly croton, hemp dogbane, wild lettuce, tall ironweed, tansy mustard, wild radish, tansy ragwort, shepardspurse, pigweeds, curly dock, galinsoga, goatsbeard, kochi, lespedeza, oxalis, pennycress, pepperweed, purslane, sneezeweed, sowthistle, Russian thistle, wild violet, vetch, wormwood, yellow rocket, bindweed, cinquefoil, dogfennel, fleabane, goldenrod, kudzu, milkweed, pokeweed, hemp sesbani, bull thistle, Canada thistle, musk thistle, yarrow.

CURTAIL (clopyralid/2,4-D)   Mfg: DowElanco

Burdock, Canadian thistle, chamomile, cocklebur, cornflower, cudweed, dandelion, flixweed, goat's beard, groundsel, hawk's beard, kochia, lambsquarters, mayweed, mustard, nightshade, pennycress, plantain, pigweed, ragweed, Russian thistle, shepard's purse, smartweed, sow thistle, sunflower, vetch, wild buckwheat, wild radish.

CURTAIL-M (clopyralid/MCPA)   Mfg: DowElanco

Alfalfa, burdock, buffalo bur, Canada thistle, chamomile, clovers, cocklebur, coffeeweed, cornflower, curly dock, dandelion, dogfennel, field pennycress, flixweed, giant ragweed, groundsel, hawksbeard, horseweed, Jerusalem artichoke, jimsonweed, knapweed, kochia, ladysthumb, lambsquarters, lentils, musk thistle, marshelder, mustards, nightshade, peas, pigweed, pineapple weed, plaintain, prickly lettuce, radish, ragweed, red sorrel, Russian thistle, salsify, shepards purse, sicklepod, smartweed, sowthistle, starthistle, sunflower, tansy mustard, velvetleaf, vetch, volunteer beans, wild buckwheat.

CYANAZINE, BLADEX   Mfg: DuPont/Griffin

Annual bluegrass, annual morningglory, annual ryegrass, annual sedge, barnyardgrass, buffalobur, bullgrass, buttercup, carpetweed, cheatgrass, chickweed, cocklebur, corn spurry, crabgrass, curly dock, downybrome, fall panicum, false flax, fescue, Flora's paintbrush, Florida pusley, foxtails, galinsoga, goosegrass, groundcherry, groundsel, henbit, Indian lovegrass, jimsonweed, Johnson grass, jungle rice, knotweed, kochia, lady's thumb, lambsquarters, mallow, mayweed, mustard, pigweed, pineapple weed, plantain, poorjoe, prickly lettuce, purslane, ragweed, Russian thistle, shepherd's purse, smartweed, spinysida, spurge, stinkgrass, tarweed, teaweed, velvetleaf, volunteer wheat, wild buckwheat, wild turnip, wild radish, wild oats, witchgrass.

CYCLE (metolochlor/cyanazine)　　　　　　Mfg: Ciba

Barnyardgrass, crabgrass, crowfoot grass, fall panicum, foxtail millet, foxtails, goosegrass, prairie cupgrass, red rice, signalgrass, southwest cupgrass, witchgrass, yellow nutsedge, nightshade, carpetweed, annual ragweed, Florida pusley, galinsoga, lambsquarters, pigweed, smartweed, prickly sida, sunflower.

CYCLOATE, RO-NEET　　　　　　Mfg: Zeneca

Alligator weed, annual bluegrass, annual rye, black nightshade, crabgrass, foxtails, goosefoot, hairy nightshade, henbit, lambsquarters, nettle, purple nutgrass, purslane, red root pigweed, shepherd's purse, stinging nettle, volunteer barley, watergrass, wild oats, yellow nutgrass, velvet leaf.

DAKOTA (fenoxaprop-ethyl/MCPA)　　　　　　Mfg: Agr Evo

Foxtails, foxtail millet, sowthistle, lambsquarters, field pennycress, pigweed, purslane, ragweed, mustards, wild radish.

DCPA, DACTHAL　　　　　　Mfg: ISK Biosciences

Annual bluegrass, barnyardgrass, carpetweed, chickweed, crabgrass, creeping speedwell, spurge, dodder, Florida pusley, foxtail, goosegrass, groundcherry, Johnson grass (from seed), lambsquarters, lovegrass, pigweed, purslane, sandbur, spurge, witchgrass, burning nettle, field pansy, copperleaf, deadnettle, knotweed, nightshade.

DERBY (metolachlor/simazine)　　　　　　Mfg: Ciba

Annual bluegrass, barnyardgrass, crabgrass, crowfootgrass, fall panicum, foxtail millet, foxtails, goosegrass, sprangletop, cupgrass, red rice, signalgrass, witchgrass, yellow nutsedge, carpetweed, chickweed, evening primrose, pigweed, spotted spurge.

DESMEDIPHAN, BETANEX　　　　　　Mfg: Agr Evo

Chickweed, coast fiddleneck, goosefoot, groundcherry, kochia, lambsquarters, London rocket, mustard, nightshade, pigweed, purslane, ragweed, shepherd's purse, sow thistle, wild buckwheat.

DETAIL (imazaquin/dimethenamid)   Mfg: American Cyanamid

Alligatorweed, beggarweed, bristly starbur, carpetweed, bur cucumber, cocklebur, hop horn bean copperleaf, jimsonweed, lambsquarter, Venice mallow, Mexicanweed, morninglory, mustard, nightshade, pigweed, wild poinsettia, puncture vine, purslane, Florida pusley, ragweed, redweed, sesbania, sicklepod, prickly sida, smartweed, spurge, sunflower, Texasweed, velvetleaf, barnyardgrass, volunteer corn, crabgrass, cupgrass, goosegrass, johnsongrass, foxtails, panicums, red rice, sheltercane, signalgrass, witchgrass, rice flat sedge.

DICAMBA, BANVEL, CLARITY,   Mfg: Sandoz
VANQUISH

Alfalfa, alligator weed, annual morningglory, arrowhead, ash, aspen, asters, balloonvine, basswood, bedstraw, beggarweed, bindweed, bitter butterweed, bitter dock, blackberry, bladder champion, bloodweed, blueball, blueweed, bracken fern, bracted plantain, broomweed, buckrush, buffalo bur, bullnettle, bur clover, bur cucumber, bursage hopclover, buttercup, Canada thistle, Carolina geranium, carpetweed, cattail, cedar, cheatgrass, cherry, chickweed, chickory geranium, carpetweed, cattail, cedar, cheatgrass, cherry, chickweed, chickory, chinquapin, clovers, cocklebur, common burdock, corn chamomile, corn cockle, cottonwood, cow cockle, crabgrass, creosote bush, croton, cucumber tree, curlydock, dalmation toad flax, dandelion, dewberry, dock, dog fennel, dogwood, dragonhead, Eastern persimmon, elm, English daisy, evening primrose, fiddleneck, field bindweed, field peppergrass, field pennycress, fleabane, Florida pusly, foxtails, french weed, frongbit, giant ragweed, goatsbeard, goldenrod, goosefoot, grape, gromwells, groundsel, hardwood trees, hawkweed, hemp dogbane, hempnettle, henbit, hickory, honeysuckle, hop clover, hornbean, horsemint, horsenettle, horseweed, Jerusalem artichoke, jimsonweed, knapweed, knawel, knotweed, kochia, kudza, lady's thumb, lambsquarters, lawn burweed, leafy spurge, locust, London rocket, lupines, mallow, maples, mare's tail, marsh thistle, mayweed, milkweeds, morningglory, multiflora rose, musk thistles, mustards, nettles, nightflowering catchfly, nightshade, nutgrass, oak, parrotfeather, pepperweed, persimmon, pickerelweed, pigweed, pine, pingue, plaintains, poison ivy, poison oak, pokeweed, poorjoe, popular, poverty weed, prickly lettuce, prickly sida, puncturevine, purslane, rabbit brush, rag-weed, rattail fescue, rattlebush, red sorrel, red vine, ripgut brome, rough stumpweed, Russian knapweed, Russian thistle, sagebrush, sassafras, service berry, sesbania, sheep sorrel, shepherd's purse, silverleaf night-shade, slender spike rush, small leafsida, smartweed, snakeweed, sneezeweed, sourwood, sow thistle, Spanish nettle, spicebush, spikeweed, spiny amaranthus, spurge spurry, starbur, star thistle, stitchwort, stinkweed, sumac, sumpweed, sunflower, swamp smartweed, sweet clover, sycamore, Tansy ragwort, tarbrush, tarweed, teasel, Texas blueweed, thistles, thorn apple, thorn berry, tievine, trumpet creeper, velvetleaf, Venice mellow, vetch, watergrass, water hemlock, water hemp, water hyacinth, water pennywort, water primrose, white cockle, wild buckwheat, wild carrot, wild cucumber, wild garlic, wild onion, wild radish, willow, witch hazel, wood sorrel, worm wood, yankeeweed, yarrow, yaupon, yucca.

DICHLORBENIL, CASORON, NOROSAC

Mfg: PBI Gordon & Uniroyal

Annual bluegrass, artemisia, bindweed, bluegrass, bull thistle, camphorweed, Canada thistle, carpetweed, chara, chickweed, citronmelon, coffeeweed, coontail, crabgrass, cudweed, dandelion, dock, dog fennel, elodea, evening primrose, false dandelion, fescue, fiddleneck, Florida purslane, foxtail, gisekia, goosefoot, groundsel, henbit, horsetail, knotweed, lambsquarters, leafy spurge, maypops, milkweed vine, minerslettuce, naiads, natalgrass, old witchgrass, orchardgrass, peppergrass, pigweed, pineappleweed, plaintain, pond weeds, purslane, quackgrass, ragweed, red deadnettle, redroot, rosary pea, Russian knapweed, Russian thistle, sheep sorrel, shepherd's purse, smartweed, Spanish needles, spurge, teaweed, Texas panicum (Harrahgrass), timothy, water milfoil, wild artichoke, wild aster, wild barley, wild carrot, wild mustard, wild radish, yellow rocket, yellow wood sorrel, hydrilla, alligator weed, watermilfoil, splatter dock, white waterlily, eelgrass, bladderwort, cattails, purple loosestrife, widgeongrass, Florida betony, red sorrel, foxtail barley, witchgrass, Virginia pepperweed, deadnettle, stranglervine, ferns rush, dodder, velvetgrass, rushes, rattlesnake grass, rice cutgrass, muskratgrass, stargrass, needlegrass, onion grass, beggarticks, tideland clover, sorrell, buckbean, hawkweed, marsh pea, pennywort, water primrose, asters.

DICLOFOP-METHYL, HOELON, ILLOXAN

Mfg: Agr Evo

Annual ryegrass, barnyard grass, broadleaf signalgrass, canarygrass, crabgrass, crowfoot grass, downybrome, fall panicum, foxtails, goosegrass, itchgrass, Persian darnel, ripgut brome, volunteer corn, wild oats, witchgrass.

DIFENZOQUAT, AVENGE

Mfg: American Cyanamid

Wild oats.

DIMETHENAMID, FRONTIER　　　　　　Mfg: Sandoz

Barnyardgrass, broadleaf signalgrass, crabgrass, foxtails, goosegrass, johnsongrass, fall panicum, cupgrass, Texas panicum, red rice, carpetweed, purslane, Florida pusley, nightshade, pigweed, spurges, rice flatsedge, yellow nutsedge.

DIQUAT, REWARD　　　　　　Mfg: Zeneca

Algae, bladderwort, cattails, coontails, Elodea, most small annual weeds, naiad, pennywort, pond weeds, salvinia, water hyacinth, water lettuce, water milfoil.

DITHIOPYR, DIMENSION　　　　　　Mfg: Rohm and Haas

Crabgrass, foxtails, barnyardgrass, crowfootgrass, lespedeza, purslane, corn speedwell, kukuyugrass, oxalis, smutgrass.

DIURON, KARMEX　　　　　　Mfg: DuPont and others

Ageratum, Amsineka (fiddleneck), annual groundcherry, annual lovegrass, annual morningglory, annual ryegrass, annual smartweed, annual sow thistle, annual sweet vernalgrass, barnyard grass, bluegrass, buttonweed, chickweed, crabgrass, corn speedwell, corn spurry, day flower, dog fennel, foxtails, gromwell, groundsel, hawkbeard, horseweed, Johnson grass (from seed), knawel, kochia, Kyllinga, lambsquarters, marigold, Mexican clover, mustard, orchardgrass, pennycress, peppergrass, pigweed, pineapple weed, pokeweed, purslane, rabbit tobacco, ragweed, rattail fescue, red sprangletop, ricegrass, sandbur, shepherd's purse, Spanish needles, tansy mustard, velvetgrass, wild buckwheat, wild lettuce, wild mustard, wild radish.

DSMA　　　　　　Mfg: ISK Biosciences and others

Bahiagrass, barnyardgrass, chickweed, cocklebur, dallisgrass, goosegrass, Johnsongrass, nutgrass, puncturevine, ragweed, sandbur, wood sorrel.

EDGE (fonofos/pebulate)　　　　　　Mfg: Zeneca

Barnyardgrass, bermudagrass, crabgrass, foxtails, goosegrass, millet, signalgrass, wild oats, blackeye susan, lambsquarters, purslane, henbit, Florida pursley, nightshade, goosefoot, pigweed, shepards purse, nutsedge, cutworms, wireworm and flea beetles.

ENDOTHAL, AQUATHOL,   Mfg: Elf Atochem
HYDROTHOL, HERBICIDE 273,
ACCELERATE

Algae, annual bluegrass, barnyardgrass, bassweed, bromes, black medic, bullgrass, bur clover, bur reed, carrotweed, cheatgrass, chickweed, clovers, coontail, cranesbill, dichrondra, foxtail, filaree, goathead, henbit, hydrilla, knotweed, kochia, little barley, lespedeza, naiad, oxalis, pigweed, pond weeds, purslane, ragweed, rescuegrass, ryegrass, shepherd's purse, smartweed, Texas blueweed, vetch, volunteer barley, veronica, watergrass, water milfoil, water stargrass, widgeongrass, wild buckwheat.

EPTC, EPTAM,   Mfg: Zeneca
ERADICANE

Annual bluegrass, annual morningglory, annual ryegrass, barnyardgrass, bermudagrass, black nightshade, carpetweed, chickweed, corn spurry, crabgrass, fiddleneck, Florida purslane, foxtail, goosegrass, hairy nightshade, henbit, Johnson grass, lambsquarters, lovegrass, fall panicum, mugwort, nettleleaved goosefoot, prostrage pigweed, purple nutgrass, purslane, quackgrass, ragweed, red root pigweed, rescuegrass, sandbur, shattercane, shepherd's purse, signal grass, tumbling pigweed, volunteer grains, watergrass, wild oats, yellow nutgrass.

ETHALFURALIN, SONALAN,   Mfg: DowElanco
CURBIT   and Platte Chemical

Annual bluegrass, barnyardgrass, carpetgrass, catchfly, chickweed, crabgrass, fall panicum, fiddleneck, Florida pusley, foxtails, foxtail millet, goosegrass, groundcherry, henbit, Johnsongrass, jungle rice, kochia, lambsquarters, nightshade, pigweed, purslane, ragweed, rock purslane, Russian thistle, ryegrass, shattercane, signalgrass, Texas panicum wild buckwheat, wild mustard, wild oats, witchgrass.

ETHOFUMESATE, NORTRON,   Mfg: Agr Evo
PROGRASS

Annual bluegrass, barnyardgrass, canarygrass, chickweed, crabgrass, downybrome, fescue, foxtails, kochia, lady's thumb, lambsquarters, mannagrass, nightshade, pigweed, puncturevine, purslane, Russian thistle, smartweed, softchess, velvetgrass, volunteer barley, volunteer wheat, wild buckwheat, wild oats.

98

EXTRAZINE II (atrazine/cyanazine)   Mfg: DuPont

Annual bluegrass, annual ryegrass, annual sedge, barnyardgrass, buffalobur, bullgrass, buttercup,carpetweed, chickweed, cocklebur, corn spurry, crabgrass, curly dock, fall panicum, fescues, fiddleneck, Florida pursley, foxtails, goosegrass, groundcherry, groundsel, jimsonweed, junglerice, kochia, ladysthumb lambsquarters, mallow, mayweed, morningglory, mustards, nightshade, pigweed, pine appleweed, plaintain, poorjoe, prickly sida, prostrate spurge, purslane, ragweed, Russian thistle, shepards purse, smallflower galinsoga, smartweed, spiny sida, stinkgrass, sunflower, tarweed, velvetleaf, wild buckwheat, wild radish, wild turnip, witchgrass.

FALLOW MASTER (glyphosate/   Mfg: Monsanto
dicamba)

Barley, barnyardgrass, blue mustard, buffalograss, cheatgrass, downybrome, fall panicum, foxtails, goatgrass, kochia, prickly lettuce, Russian thistle, tansy mustard, tumble mustard, wheat, wild oats.

FENOXAPROP-P-ETHYL, ACCLAIM,   Mfg: Agr Evo
WHIP, HORIZON, OPTION, BUGLE

Barnyardgrass, broadleaf signalgrass, crabgrass, cupgrass, foxtails, goosegrass, itchgrass, Johnson grass, jungle rice, panicums, red rice, sandbur, sprangletop, volunteer corn, wild cane, wild oats, wild proso millet, wirestem muhly, witchgrass.

FERRIC SULFATE   Mfg: Numerous

Moss.

FINESSE (chlorsulfuron/   Mfg: DuPont
metsulfuron-methyl)

Annual bluegrass, annual ryegrass, bedstraw, broadleaf dock, buckwheat, bur breakchevil, bur buttercup, Canadian thistle, chickweed, conical catchfly, corn groundsel, corn spurry, cow cockle, dovefoot geranium, false chamomile, fiddleneck, filaree, field pennycress, flixweed, foxtails, groundsel, hempnettle, henbit, Jacobsladder, knotweed, kochia, lady's thumb, lambsquarters, little bittercress, Mayweed, miners lettuce, pigweed, pineapple weed, prickly lettuce, prickly poppy, purslane, Russian thistle, shepherd's-purse, smartweed, sowthistle, speedwell, tansy mustard, vetch, white cockle, wild buckwheat, wild mustard, wild radish.

FLUAZIFOP-P-BUTYL, FUSILADE 2000   Mfg: Zeneca

Barnyardgrass, bermudagrass, canarygrass, crabgrass, cupgrass, downy brome, fall panicum, foxtails, goosegrass, itchgrass, Johnson grass, jungle rice, quackgrass, red rice, sandbur, shattercane, signal cane, signal grass, Texas panicum, volunteer cereals, volunteer corn, wild oats, wild proso millet, wirestem muley, witchgrass.

FLUOMETURON, COTORAN,   Mfg: CIBA
METURON

Barnyardgrass, Brachiaria, buttonweed, cocklebur, crabgrass, crowfoot grass, fall panicum, Florida pusley, foxtail, goosegrass, groundcherry, jimsonweed, lambsquarters, morningglory, pigweed, prickly sida, puncturevine, purslane, ragweed, sesbania, ryegrass, sicklepod, smartweed, tumbleweed.

FLURIDONE, SONAR   Mfg:Sepro

Bladderwort, common coontail, common duckweed, creeping waterprimrose, egeria, elodea, hydrilla, naiad, paragrass, pondweed, reed canarygrass, spatterdock, water lily, water milford, water purslane.

FOMESAFEN, REFLEX   Mfg: Zeneca

Amamranth, bristly starbur, carpetweed, citron, cocklebur, copperleaf, crotalaria, eclipta, Florida pusley, giant ragweed, hemp sesbania, hop hornbeam, jimson-weed, lady's thumb, lambsquarters, Mexican weed, morningglory, prickly sida, pruslane, ragweed, red weed sicklepod, smartweed, smell melon, spurge, spurred anoda, tall waterhemp, tropic croton, velvetleaf, Venice mallow, Virginia copperleaf, wild cucumber, wild mustard, willowleaf, witchgrass, yellow rocket.

FOSAMINE, KRENITE   Mfg: DuPont

Alder, ash, black gum, blackberry, cottonwood, Eastern white pine, elm, hawthorn, hickory, locust, maple, multiflora rose, oaks, pines, poplar, sassafras, sumac, sweetgum, sycamore, thimbleberry, tree of heaven, vine maple, wild cherry, wild grape, wild plum, willow.

FREEDOM (alachlor/trifluralin)   Mfg: Monsanto

Barnyardgrass, broadleaf signalgrass, carpetweed, crabgrass, Florida pursly, foxtails, galinsoga, goosegrass, groundcherry, nightshade, panicums, pigweed, purslane, red rice, sprangletop, waterhemp, witchgrass, yellow nutsedge.

FUSION, HORIZON 2000
(fenoxaprop-p-ethyl/fluazifop-p-butyl)　　　Mfg: Zeneca/Agr Evo

Barnyardgrass, broadleaf signalgrass, downy brone, crabgrass, fall panicum, fescues, sandbur, foxtails, goosegrass, itchgrass, seedling johnsongrass, junglerice, quackgrass, red rice, ryegrass, shattercane, southwestern cutgrass, Texas panicum, volunteer cereals, wild oats, wild prose millet, wirestem mukly, witchgrass, woolly cupgrass.

GALAXY (acifluorfen-sodium/　　　Mfg: BASF
bentazon)

Beggarticks, bristly starbur, Canadian thistle, cocklebur, common ragweed, day flower, devils claw, galinsoga, giant ragweed, jimsonweed, lady's thumb, lambsquarters, morningglory, pigweed, prickly sida, purslane, redweed, shepard's purse, smartweed, spurred anoda, sunflower, velvetleaf, Venice mallow, wild buckwheat, wild mustard, wild poinsetta, yellow nutsedge.

GEMNI (chlorimuron-ethyl/linuron)　　　Mfg: DuPont

Bur cucumber, carpetweed, cocklebur, common ragweed, Florida beggarweed, giant ragweed, hemp jimsonweed, hemp sesbania, lambsquarters, morningglory, mustards, nutsedge, pigweed, purslane, sesbania, sicklepod, smartweed, sunflower, teaweed velvetleaf.

***GLUFOSINATE-AMMONIUM*, IGNITE,**　　　Mfg: Agr Evo
FINALE

Chickweed, clover, cocklebur, filaree, jimsonweed, kochia, spurge, London rocket, malva, maristail, purslane, shepardspurse, smartweed, barnyardgrass, cupgrass, fall panicum, foxtails, goosegrass, johnsongrass, lovegrass, shattercase, stinkgrass, windgrass, buffalobur, Canada thistle, curly dock, dogbane, field gromwell, fleabane, lambsquarters, mush thistly nightshade, pennycress, pigweed, prickly lettuce, ragweed, tansy mustard, velvetleaf, Virginia copperleaf, aster, wild buckwheat, mustards, wild onion, wild rose, wild turnip, annual bluegrass, barley, crabgrass, downy brome, Kentucky bluegrass, sandbur, bromes, wheat, wild oats, sowthistle, bindweed, burdock, dandelion, goldenrod, horsetail, nettle, plaintain, Russian thistle, vervain, wood sorrel, yellow rocket, bahiagrass, Bermudagrass, carpetgrass, dallisgrass, fescue, guiniagrass, nutsedge, paragrass, quackgrass, ryegrass, torpedograss, vaseygrass.

GLYPHOSATE, RODEO, JURY, ROUNDUP, ACCORD, PROTOCOL, RANGER, GLYFOS, HONCHO

Mfg: Monsanto/Cheminova Agro

Alder, alfalfa, alligator weed, annual bluegrass, ash, aspen, bahiagrass, balsam apple, barley, barnyard grass, bassia, bearclover, bearmat, bedstraw, bermudagrass, berries, bindweed, birch, blackberry, bluegrass, bracken fern, Brazilian peppertree, brome, bulbous bluegrass, buttercup, California buckwheat, camelthorne, Canadian thistle, canarygrass, Carolina foxtail, Carolina geranium, cascara, cats claw, cattails, ceanothis, chamise, cheatgrass, cherry chervil, chickweed, clanothus, clovers, clocklebur, common mullein, cordgrass, coyote brush, crabgrass, curly dock, cutgrass, dallisgrass, dandelion, dewberry, dogbane, downybrome, dwarf dandelion, elderberry, elm, eucalyptus, evening primrose, falseflax, fescues, fiddleneck, field bindweed, filaree, flaxleaf fleabane, Florida pusley, foxtail, French broom, giant cutgrass, giant reed, giant ragweed, glyphosate, groundsel, guineagrass, hasarchia, hawthorne, hazel, hemp dog-bane, henbit, holly, honeysuckle, horsenettle, horseradish, horseweed, Jerusalem artichoke, Johnson grass, Jointed goatgrass, Kentucky bluegrass, kikuyugrass, knapweed, kochia, kudzu, lambsquarters, lantana, locust, London rocket, loosestrife, lotus, maidencane, manzanita, maples, milkweed, monkey flower, morningglory, mullein, multiflora rose, mustards, Napiergrass, nutsedge, oaks, orchardgrass, panicum, paragrass, pennycress, persimmon, phragmites, pigweed, pines, poison ivy, poison oak, poplar, prickly lettuce, quackgrass, ragweed, raspberry, reed canary grass, Russian thistle, ryegrass, sage, sagebrush, salmon berry, salt brush, sandbur, sassafras, Scotch broom, shattercane, shepherd's-purse, signal grass, silver nightshade, smartweed, sourwood, sowthistle, Spanish needles, spatter dock, speedwell, spotted spurge, stinkgrass, sumac, sunflower, sweet gum, swordfern, tallow tree, teaweed, Texas blueweed, thimble berry, timothy, torpedograss, tree tobacco, trumpet creeper, tules, umbrella spurry, vaseygrass, velvetleaf, vetch, Virginia creeper, volunteer corn, volunteer wheat, water hyacinth, water lettuce, water primrose, wax myrtle, wheat, wheat grass, wild oats, wild sweet potatoes, willow, wirestem.

GLYPHOSATE-TRIMESIUM, TOUCHDOWN

Mfg: Zeneca

Annual bluegrass, chickweed, cupgrass, foxtails, jimsonweed kochia, wild oats, puncturevine, ryegrass, signal grass, sowthistle, spurge, Russian thistle, volunteer cereals, witchgrass, volunteer alfalfa, little barley, barnyardgrass, crabgrass, evernig primrose, goosegrass, lambsquarters, mustard, nightshade, panic grass, pineappleweed, quackgrass, sicklepod, smartweed, timothy downy brone, cocklebur, hemp dogbane, hophorn bean, johnsongrass, milkweed, puslane, wild radish, ragweed,shepards purs, morningglory, teaweed, Canada thisle, bermudagrass, field bindweed, Cogongrass, tall fescue yellow nutsedge, sesbaina, velvetleaf.

GUARDSMAN (dimethenamid/ atrazine)
Mfg: Sandoz Agro

Pigweeed, lambsquarters, ragweed, nightshade, morninglory, jimson-weed, barnyardgrass, broadleaf signalgrass, crabgrass, foxtails, goosegrass, johnsongrass, fall panicum, red rice, cupgrass, Texas panicum, wild oats, signalgrass, carpetweed, cocklebur, purslane, Florida pusley, jimsonweed, kochia, lambsquarter, morninglory, mustards, night-shade, Palmer anaraanth, pigweed, ragweed, smartweed, spurge, tall waterbug, velvetleaf, wild buckwheat, rice flatsedge, yellow nutsedge.

HARMONY EXTRA (trifensulfuron methyl/tribenuron-methyl)
Mfg: DuPont

Annual knawel, annual sow thistle, mustards, broadleaf dock, bur butter-cup, busky wallflower, pepperweed, fiddleneck, buckwheat, chickweed, groundsel, lambsquarters, wild radish, sunflower, corn chamomile, corn gromwell, corn spurry, cow cockle, curly dock, field pennycress, false chamomile, flaxweed, smartweed, henbit, kochia, ladysthumb, London rocket, marsh elder, mayweed, miners lettuce, mouse ear cress, nightflowering catchfly, pine appleweed, prickly lettuce, knotweed, pig-weed, redmaids, Russian thistle, shepardspurse, buttercup, smallseed false flax, dog fennel, swinecress, tansy mustard, tarweed, volunteer lentils, volunteer peas, wild buckwheat, wild chamomile, wild garlic.

HEXAZINONE, VELPAR
Mfg: DuPont

Annual bluegrass, annual ryegrass, bahiagrass, barnyard grass, bermudagrass, bindweed, bluegrass, bouncing bet, broomsedge, bro-megrass, burdock, camphorweed, Canada thistle, cheatgrass, chick-weed, clovers, cocklebur, crabgrass, crown vetch, dallisgrass, dande-lion, deerbrush, dewberry, dock, dog fennel, dogbane, fescue, fiddleneck, field pennycress, filaree, fingergrass, fleabane, foxtail, foxtail barley, goatsbeard vine, goldenrod, groundsel, guinea grass, heath aster, henbit, honeysuckle, ivyleaf speedwell, kochia, lambsquarters, lantana, lespe-deza, London rocket, manzanita, marestail, Mexican tea, milkweed, miners lettuce, mustards, notchgrass, nutsedge, orchardgrass, oxalia, paragrass, pigweed, plaintain, prickly lettuce, purslane, quackgrass, ragweed, ryegrass, seedling alfalfa, shepherd's-purse, smartweed, smutgrass, Spanish needle, spurge, star thistle, sweet clover, tansy mustard, trumpet creeper, vaseygrass, wild blackberry, wild carrot, wild oats, wild parsnip, yellow rocket,English catchfly, salesify, spurry, wild radish, speedwell, white cockle, ageratim, Bakan apple, castorbean, crotalari, jungle rice, oxalis, popolo, vasey grass, alerandergrass, amaranth, fireweed, Carolina ger, Flora's paintbrush, johnsongrass, sandbur, signalgrass, sowthistle, sunflower, sensitive plant, waltheria, aster, bracken fern, horseweed, fireweed, panicum, quackgrass, ash, aspen, birch, Balsam poplar, birch, box elder, branbles, cherry dogwood, elm, hawthorne, hazel, honesuckle, oaks, maple, sweetgum, willow, wourwood, bentgrass, elksedge, fleabane, oxyeye darsy, pinegrass, velvetgrass, catsear, snowbrush, manzanita, squaw carpet, deerbrush, blackgum, hickory, persimmon, hackberry, locust, mesquite, mulbury, sweetbug.

IMAZAMETHABENZ-METHYL,   Mfg: American Cyanamid
ASSERT

Bedstraw, catchweed, field pennycress, flixweed, London rocket, shepard's purse, tansy mustard, wild buckwheat, wild mustard, wild oats, wild radish.

IMAZAPYR, ARSENAL, CHOPPER   Mfg: American Cyanamid

Annual bluegrass, ash, bahiagrass, beardgrass, bermudagrass, big bluestem, bindweed, blackberry, bluegrass, boxelder, broadleaf signalgrass, bull thistle, burdock, camphorweed, Canada bluegrass, Canada thistle, canarygrass, Carolina geranium, carpetweed, cattail, cheat, cherry, chickweed, Chinese tallow-tree, churchgrass, clover, cocklebur, crabgrass, curly dock, dallisgrass, dandelion, dewberry, dogfennel, dogweed, dogwood, downybrome, feather top, fescue, filaree, fleabane, foxtails, goldenrod, goosefoot, goosegrass, greenbriar, guineagrass, hawthorne, hickory, hoary vervain, honey locust, honeysuckle, Johnson grass, Kentucky bluegrass, kochia, kudzu, lambsquarters, lespedeza, lovegrass, mallow, maple, marestail, milkweed, miner's lettuce, morningglory, mulberry, mullein, multiflora rose, mustard, oak, opossum grape, orchardgrass, ox-eye daisy, panicum, paragrass, pepperweed, persimmons, phargonites, pigweed, plaintain, poison ivy, pokeweed, popular, prairie cordgrass, prairie threeawn, primrose, privet, purslane, quackgrass, ragweed, redvine, ryegrass, salt cedar, saltgrass, sandbur, sand dropseed, sassafras, signalgrass, silverleaf nightshade, smartweed, smooth brome, sorrel, sow thiatle, sumac, sweet clover, sweetgum, tall fescue, Texas thistle, timothy, torpedograss, trumpet creeper, vaseygrass, Virginia creeper, wild barley, wild buckwheat, wild carrot, wild grape, wild lettuce, wild mustard, wild oats, wild parsnip, wild turnip, willow, wirestem muhly, witchgrass, willyleaf bursage, yellow star thistle, yellow poplar, yellow wood sorrel, aspen, olive, boxelder, cottonwood, alder, Russian olive, tulip poplar, beech, aspen, cypress, blackgrass, chinaberry, alder, sourwood, sweetgum, rose snakeweed, desert camelthorn, knapweed, skeleton weed, nettle, Japanese bamboo, sowthistle, Russian olive, salt cedar, rabbitbrush, bedstraw, bishopweed, buttercups, little barley, wood sorrel.

IMAZAQUIN, SCEPTER, IMAGE   Mfg: American Cyanamid

Annual bluegrass, alligator weed, burweed, chickweed, cocklebur, dollarweed, Florida pusley, foxtails, giant ragweed, globe sedge, green kyllinga, henbit, jimsonweed, lambsquarters, morningglory, mustard, nightshade, nutsedge, parsley-piert, pigweed, purple nutsedge, rice flatsedge, ryegrass, sandbur, ragweed, rice flatsedge, seedling Johnson grass, smartweed, teaweed, torpedograss, velvetleaf, Venice mallow, wild garlic, wild onion, wild poinsettia, wild sunflower, field sandbur, black medic, buttercup, evening primrose, geranium, hairy bittercress, knauvel, lawn burweed, parsley-piert, purple dead nettle, red sorrel, white clover, cudweed, dandelion, annual bluegrass.

IMAZETHAPYR, PURSUIT   Mfg: American Cyanamid

Annual ragweed, alligator weed, barnyardgrass, broadleaf signalgrass, carpetweed, cocklebur, crabgrass, Florida pusley, foxtails, galissoga, giant ragweed, Jerusalem artichoke, jimsonweed, Johnsongrass, kochia, lambsquarters, morningglory, mustards, nightshade, panicums, pigweed, puncture vine, purslane, shattercane, smallflower morningglory, smartweed, spotted spurge, tall waterhemp, teaweed, Texas panicum, velvetleaf, wild sunflower, wooly cupgrass.

ISOXABEN, GALLERY   Mfg: DowElanco

Bittercress, brassbuttons, Carolina geranium, chickweed, cudweed, fiddleneck, filaree, fleabane, groundsel, henbit, horseweed, knotweed, ladysthumb, lambsquarters, London rocket, mustards, nightshade, pennywort, pigweed, pineappleweed, plaintains, prickly lettuce, purslane, ragweed, redmaids, shepards purse, sibara, sowthistle, speedwell, spurge, tansy mustard, velvetleaf, white clover, wild celery, wood sorrel, dog fennel, galensoga, dandelion, mallow, aster, buroage, clovers, pepperweed, wild celery, ground cherry, Russian thistle, wild radish, wild carrot, smartweed, telegraph plant, chamberbitter, marestail, prayweed, morninglory, phylianthus, pokeweed, prickly sida, red sorrel, sweet cloves, yellow wood sorrel, bur clover, datura, dog fennel, evening primrose, fescues, goosefoot, jimsonweed, aster, bursage, wild celery, clover, ground cherry, mallow, pepperweed, London rocket, smartweed, telegraph plant, Russian thistle, bittercress, brassbuttons, wild carrot, chamberbitter, Carolina geranium, ecipta, marestail, mayweed, morninglory, pennywort, pokeweed, prickly sida, red sorrel, sweet clover, tansy mustard, bur clover, dog fennel, evening primrose, fescue, jimsonweed, knotweed, goosefoot, kochia, black media, nettle, bristly ox tongue, scarlet pimpernel, sunflower, swinecress, mush thistle, willow wwed, wood sorrel, bindweed, carpetweed, curly dock, venue mallow, milkweed, Florida prusley.

KOVAR (bromacil/diuron)   Mfg: DuPont

Bahiagrass, barnyard grass, bermudagrass, bouncing bet, chickweed, crabgrass, dogbane, filaree, fleabane, Florida pusley, foxtails, groundsel, horseweed, Johnson grass, jungle rice, lambsquarters, natalgrass, nightshade, nutsedge, pangolagrass, pigweed, paragrass, pineapple weed, puncturevine, purslane, ragweed, saltgrass, sandbur, seedling sow thistle, Spanish needles, torpedograss, wild lettuce, wild mustard.

LACTOFEN, COBRA   Mfg: Valent

Balloon vine, beggarticks, bristly starbur, buffalobur, bur cucumber, carpetweed, cocklebur, common ragweed, copperleaf, devil's claw, eclipta, Florida beggarweed, Florida pusley, giant ragweed, groundcherry, galinsoga, hemp sesbania, jimsonweed, morningglory, nightshade, pigweeds, poorjoe, prickly sida, purslane, showy crotalaria, smell melon, spurges, tall water hemp, tropic croton, velvetleaf, Venice mallow, wild mustard, wild poinsettia, wild sunflower, witchweed, mexicanweed, puncturevine, ragweed.

LADDOK, PROMPT   Mfg: BASF
(bentazon/atrazine)

Beggarticks, bristly starbur, Canada thistle, cocklebur, dayflower, field bindweed, giant ragweed, jimsonweed, kochia, lady's thumb, lambsquarters, morningglory, pigweed, prickly sida, small flower morningglory, smartweed, spurred anoda, starbur, sunflower, tall waterhemp, velvetleaf, Venice mallow, wild buckwheat, wild mustard, yellow nutsedge.

LANDMASTER BW   Mfg: Monsanto
(glyphosate/2,4-D)

Barley, barnyardgrass, blue mustard, buffalograss, cheatgrass, cocklebur, downy brome, fall panicum, field bindweed, foxtails, goatgrass, kochia, lambsquarters, pigweed, prickly lettuce, puncture vine, purslane, Russian thistle, rye, stinkgrass, tansy mustard, tumble mustard, wheat, wild oats, witchgrass.

LARIAT (alachlor/atrazine)   Mfg: Monsanto

Annual ragweed, barnyardgrass, beggarweed, carpetweed, cocklebur, crabgrass, fall panicum, Florida pusly, foxtails, goosegrass, grassbur, jimsonweed, kochia, lambsquarters, morningglory, mustards, nightshade, pigweed, purslane, red rice, sandbur, seedling Johnsongrass, signalgrass, smartweed, teaweed, velvetleaf, witchgrass, yellow nutsedge.

LINURON, LOROX, LINEX   Mfg: DuPont

Amsinckia, annual ryegrass, buttonweed, canarygrass, carpetweed, chickweed, crabgrass, dog fennel, fall panicum, Florida pusley, foxtails, galinsoga, goosefoot, goosegrass, gromwell, groundsel, knawel, lambsquarters, morningglory, mustard, nettleleaf, pigweed, prickly sida, purslane, ragweed, rattail fescue, sesbania, sicklepod, smartweed, Texas panicum, velvetleaf, watergrass, wild buckwheat, wild radish.

LOROX PLUS (chlorimuron/linuron)   Mfg: DuPont

Annual ragweed, carpetweed, cocklebur, jimsonweed, lambsquarters, mustards, pigweed, prickly sida, purslane, smartweed, spotted spurge, tall waterhemp, velvetleaf.

MCPA, RHOMINE   Mfg: Numerous

Arrowhead, beggarsticks, bullrush, burhead, burdock, buttercups, Canada thistle, cocklebur, curly indigo, dandelion, dragonhead mint, goatsbeard, hempnettle, kochia, lambsquarters, marsh elder, meadow buttercup, mustards, nightshade, peppergrass, pigweed, pennycress, plaintain, poison hemlock, puncturevine, purslane, ragweed, redstem, Russian thistle, sedge, shepherd's-purse, sow thistle, stinkweed, stinging nettle, sunflowers, vetch, water plaintain, white top, wild radish, winter cress, water hyssop, yellow rocket.

MCPB, THISTOL   Mfg: Rhone Poulenc

Canada thistle, fanweed, lambsquarters, morningglory, nightshade, pigweed, smartweed, sowthistle.

MARKSMAN (dicamba/atrazine)   Mfg: Sandoz

Alfalfa, bindweed, bur cucumber, Canadian thistle, Carolina horsenettle, chickweed, clover, cocklebur, dandelion, dock, giant ragweed, hemp dogbane, Jerusalem artichoke, jimsonweed, kochia, lady's thumb, lambsquarters, lespedeza, mallow, milkweed, morningglory, mustard, nightshade, pigweed, prickly sida, puncture vine, purslane, Russian thistle, sicklepod, smartweed, Spanish needles, spurge, sunflower, tansy mustard velvetleaf, Venice mallow, vetch, wild buckwheat, wild cucumber.

MECOPROP, MCPP　　　　　　Mfg: PBI Gordon, Riverdale and others

Chickweed, cleavers, clover, dichronda, field pennycress, ground ivy, knotweed, lambsquarters, mustard, pigweed, plaintain, ragweed, shepherd's-purse, wild radish.

METOLACHLOR, DUAL, PENNANT, MEDAL　　　　Mfg: CIBA

Barnyardgrass, carpetweed, crabgrass, crowfoot grass, cupgrass, fall panicum, Florida pusley, foxtail millet, foxtails, galinsoga, goosegrass, nightshade, pigweed, red rice, shattercane, signal grass, witchgrass, yellow nutsedge, annual bluegrass, groundsel, wooly cupgrass, eclipta.

METRIBUZIN, LEXONE, SENCOR　　　　Mfg: Miles Inc. & DuPont

Ageratium, alexandergrass, alkali mallow, annual bluegrass, bedstraw, beggarweed, bindweed, buffalobur, bur clover, canarygrass, Carolina geranium, carpetweed, cheatgrass, chickweed, cocklebur, coffeeweed, conical catchfly, corn cockle, corn speedwell, crabgrass, crowfootgrass, curly dock, dayflower, deadnettle, dogfennel, euphorbia, fiddleneck, field pennycress, filaree, fireweed, Flora's paintbrush, Florida purslane, foxtails, galinsoga, goosefoot, goosegrass, gromwell, guiniagrass, Haoloe koa, henbit, Hialsa, Hilgihila, hop clover, horseweed, Jacob's ladder, jimsonweed, knotweed, lambsquarters, London rocket, madwort, malva, marestail, Mexican weed, miners lettuce, mustards, panicum, parsleypiert, pigweed, pineappleweed, plushgrass, prickly lettuce, prickly sida, purslane, rabbits foot grass, ragweed, rattlepot, rice grass, Richardia, seedling Johnsongrass, sesbania, hepherd's-purse, sicklepod, signal grass, six weeks gramagrass, smallflowered buttercup, smartweed, sowthistle, speedwell, spiny amarathus, spurweed, spurge, tarweed, toadflax, velvetleaf, venice mallow, white clover, winter oats, wiregrass, Jacobs ladder, catchfly, speedwell, bittercress, buttercup, evening primrose, kochia, pepperweed, shepards purse, wild turnip, wild radish, downy brome, little barley, rescuegrass, rye grass, white willowgrass.

METSULFURON-METHYL, ALLY,   Mfg: DuPont
ESCORT, DMC

Bahiagrass, bittercress, bur buttercup, Canadian thistle, chickweed, conical catchfly, corn cockle, corn gromwell, cowcockle, dogfennel, false chamomile, fiddleneck, field pennycress, filaree, flixweed, groundsel, henbit, knotweed, kochia, lambsquarters, Mayweed, miners lettuce, mustards, pigweed, prickly lettuce, purslane, Russian thistle, shepherd's-purse, smallseed toadflax, smartweed, sowthistle, sunflower, tansy mustard, treacle mustard, tumble mustard, waterpod, wild buckwheat, multiflora rose, blackberry, alfalfa, clover, ground ivy, ryegrass, beggarweed, dollarweed, Florida betoony, lespedeza, parsley-piert-pusley, ragweed, dog fennel, sida, spurweed, wild celery.

MOLINATE, ORDRAM   Mfg: Zeneca

Annual sedges, barnyardgrass, Brachiaria, crabgrass, dayflower, red rice, spike rush, watergrass.

MONCIDE, BROADSIDE (cacodylic   Mfg: Monterey/Drexel
acid/MSMA)

Barnyardgrass, goosegrass, crabgrass, foxtails, johnsongrass, Dallisgrass, nutsedge, sandbur, chickweed, purslane, puncturevine, cocklebur, ragweed.

MSMA, BUENO, WEED-HOE,   Mfg: ISK Biosciences
ANSAR   and others

Bahiagrass, barnyardgrass, chickweed, cocklebur, crabgrass, dallisgrass, Johnson grass, foxtails, mustard, nutgrass pigweed, puncturevine ragweed, sandbur, tules, wild oats, wood sorrel.

NAPROPAMIDE, DEVRINOL   Mfg: Zeneca

Annual bluegrass, barnyardgrass, bromes, carpetweed, cheatgrass, crabgrass, cupgrass, chickweed, fall panicum, fiddleneck, filaree, foxtails, goosegrass, groundsel, knotweed, lambsquarters, little mallow, mallow, panicum, pineappleweed, purslane, prickly lettuce, pigweed, ragweed, ripgut brome, ryegrass, sandbur, seedling Johnson grass, soft chess, sowthistle, sprangletop, stinkgrass, wild barley, wild oats, witchgrass.

NAPTALAN, ALANAP   Mfg: Uniroyal Inc.

Barnyardgrass, bindweed, carpetweed, chickweed, cocklebur, crabgrass, cupgrass, foxtails, galinsoga, goosegrass, groundcherry, Johnson grass (from seed), lambsquarters, mustard, nightshade, pigweed, purslane, ragweed, sandbur, shepherd's-purse, sprangletop, stinkgrass, velvetleaf, watergrass, windmill grass.

NICOSULFURON, ACCENT   Mfg: DuPont

Shattercane, johnsongrass, quackgrass, wooly cupgrass, foxtails, panicums, barnyardgrass, broadleaf signalgrass, sandbur, morninglory, pigweed, smartweed, jimsonweed.

NORFLURAZON, EVITAL,   Mfg: Sandoz
SOLICAM, ZORIAL, PREDICT

Annual bluegrass, annual bursage, annual ragweed, annual sedge, bahiagrass, barnyardgrass, bog rush, broomsedge, Carolina geranium, carpetweed, cheat, cheeseweed, chickweed, coffee senna, crabgrass, cudweed, cupgrass, dog fennel, downybrome, fall panicum, false dandelion, feather fingergrass, fescues, fiddleneck, filaree, fleabane, Florida pusley, foxtail barley, foxtails, goosegrass, Johnsongrass seedlings, London rocket, mayweed, needlegrass, nutgrass, panicums, pigweed, pineapple weed, prickly sida, povertygrass, puncturevine, purslane, redmaids, redroot, redtopgrass, rice cutgrass, ripgut brome, Russian thistle, ryegrass, salt grass, hepherd's-purse, sicklegrass, signalgrass, six weeks gramagrass, smartweed, smokegrass, spikerush, sprangle top, spurge, spurred anoda, sow thistle, stargrass, summergrass, switchgrass, tall fescue, tropic croton, velvetleaf, Virginia pepperweed, wild barley, wild buckwheat, wild geranium, wild onion, witchgrass, woolgrass, beggarweed, sicklepod, coffee senna.

ORNAMENTAL HERBICIDE II   Mfg: Scotts
(oxyfluorfen/pendimethalin)

Annual bluegrass, barnyardgrass, crabgrass, bitter cress, chickweed, cudweed, dandelion, fireweed, fleabane, groundsel, marestand, oxales, pearl wort, pepperweed, pegweed, shepardspurse, sowthistle, spurge.

ORYZALIN, SURFLAN   Mfg: DowElanco

Annual bluegrass, annual ryegrass, barnyardgrass, bittercress, Brachiaria, browntop panicum, carpetweed, chickweed, crabgrass, creeping woodsorrel, crowfootgrass, cupgrass, fall panicum, fiddleneck, filaree, Florida pusley, foxtails, goosegrass, groundsel, henbit, horseweed, johnson grass (from seed), jungle rice, knotweed, lambsquarters, little barley, London rocket, lovegrass, oxalis, panicums, pigweed, prostrate spurge, puncturvine, purslane, redmaids, red rice, sandbur, shepherd's-purse, sowthistle, sprangletop, spurge, wild oats, witchgrass.

OXADIAZON, RONSTAR   Mfg: Rhone Poulenc

Annual bluegrass, barnyardgrass, bittercress, carpetweed, crabgrass, common dayflower, crabgrass, duck salad, evening primrose, fall panicum, Florida pusley, galinsoga, golden ragwort, goosegrass, groundsel, junglerice, lambsquarters, liverwort, niruri, oxalis, Panama paspalum, petty spurge, pigweed, poa annua, prostrate spurge, purslane, red sprangletop, redstem, sensitive plant, smartweed, sowthistle, speedwell, spiny amaranthus, spotted catsears, spurge, stinging nettle, swine cress, Texas panicum, water hyssop, white stemed filaree, woodsorrel, yellow woodsorrel, clover, groundsel, red sandspurry, smartweed, fiddleneck, wild oats.

OXYFLUORFEN, GOAL   Mfg: Rohm and Haas

Annual bluegrass, annual morningglory, balsum apple, barnyardgrass, bedstraw, bittercress, burclover, camphorweed, canarygrass, carpetweed, cheeseweed, chickweed, clovers, cocklebur, corn spurry, crabgrass, cudweed, cupgrass, evening primrose, fall panicum, fiddleneck, filaree, fireweed, flexweed, foxtails, goosegrass, groundcherry, groundsel, hemp sesbania, henbit, horseweed, jimsonweed, knotweed, lady's thumb, lambsquarters, lanceleaf sage, London rocket, malva, mayweed, miners lettuce, mustards, nettle, nightshade, oxalis, pepperweed, pigweed, pineapple, prickly lettuce, puncturevine, purslane, red maids, red orach, red sandspurry, red sorrel, ripgut brome, Russian thistle, scarlet pimpernel, shepherd's-purse, smartweed, sow thistle, speedwell, spurge, sweet clover, tansy mustard, teaweed, tropic croton, velvetleaf, wild buckwheat, wild oats, wild poinsettia, wild radish, witchgrass, witchweed, yellow sorrel.

PARAQUAT, GRAMOXONE,   Mfg: Zeneca
STARFIRE, CYCLONE

Annual grasses, annual ryegrass, bermudagrass, bluegrass, burclover, cheatgrass, chickweed, crabgrass, filaree, groundsel, Johnson grass, morningglory, nettle, pigweed, plaintain, puncturevine, purslane, red clover, shepherd's-purse, thistles, volunteer barley, wild mustard, wild oats, wild radish, maristail, prickly lettuce.

PASSPORT (imazethapyr/trifluralin)   **Mfg: American Cyanamid**

Barnyardgrass, crabgrass, crowfootgrass, cupgrass, foxtails, goosegrass, seedling johnsongrass, panicums, sandbur, shattercane, broadleaf signalgrass, witchgrass, carpetweed, cocklebur, galinsoga, jimsonweed, morningglory, mustards, nightshade, pigweed, puncturevine, purslane, Florida pusley, prickly sida, smartweed, spurge, sunflower, velvetleaf, tall waterhemp.

PATHWAY (picloram/triclopyr)   **Mfg: DowElanco**

Ailanthus, alder, ash, aspen, birch, cedar, cherry dogwood, elm, firs, gum hawthorne, hickory, hornbean, maples, oaks, pecans, persimmon, serviceberry, sourwood, sweetbay.

PEBULATE, TILLAM   **Mfg: Zeneca**

Barnyardgrass, bermudagrass, blackeyed susan, crabgrass, deadnettle, Florida pusley, foxtails, goosefoot, goosegrass, hairy nightshade, lambsquarters, pigweed, purple nutgrass, purslane, shepherd's-purse, signalgrass, wild oats, yellow nutgrass.

PELARGONIC ACID,   **Mfg: Mycogen**
SCYTHE

Mosses, algae, lichens, liverworts, annual weeds, perennial weeds.

PENDIMETHALIN, PROWL   **Mfg: American Cyanamid**
STOMP, PENDULUM

Annual bluegrass, annual spurge, barnyardgrass, burweed, carpetweed, chickweed, corn speedwell, crabgrass, crowfootgrass, fiddleneck, filaree, Florida pusley, foxtails, goosegrass, henbit, itchgrass, johnsongrass seedlings, junglerice, knotweed, kochia, lambsquarters, London rocket, lovegrass, panicums, pigweed, proso millet, puncturevine, purslane, shepherd's-purse, signalgrass, smartweed, sprangletop, spurge, velvetleaf, witchgrass, wolly cupgrass, cudweed, sandbur, yellow woodsonel.

PHENMEDIPHAM, SPIN-AID   **Mfg: Agr Evo**

Chickweed, fiddleneck, goosefoot, green foxtail, groundcherry, kochia, lambsquarters, London rocket, mustard, nightshade, pigeongrass (yellow foxtail), pigweed, purslane, ragweed, shepherd's-purse, sow thistle, wild buckwheat.

PICLORAM, TORDON Mfg: DowElanco

Ailanthus, alder, artichoke thistle, ash, aspen, balsam, bindweed, birch, bitterweed, blackgum, blackberry, bouncing bet, brackenfern, brambles, broomweed, bursage, button brush, bur ragweed, Canadian thistles, cedar, chicory, chokeberry, cholla cacti, conifers, cypress, dandelion, dock, dogwood, elderberry,elm, field bindweed, fir, fleabane, goldenrod, golfberry, grose, groundcherry, gums, hawthorn, hemlock, hickory, honeysuckle, horse nettle, juniper, knapweed, kudzu, larkspur, leafy spurge, locoweeds, locusts, lupines, manzanita, maple, mesquite, milkweed, most broadleaf weeds, mulberry, musk thistle, oak persimmon, pigweed, pine, plaintain, poison oak, poverty weed, prickly lettuce, prickly pear, rabbit brush, ragweed, redband, Russian knapweed, sassafras, scotch thistle, skeleton weed, snakeweed, sourwood, sow thistle, spruce, starthistle, sumac sunflower, sweetclover, Tansy ragwort, tansy thistle, toadflax, tulip poplar, vetch, wild carrot, wild cherry, wild grapes, wild rose, willows.

PREVIEW (chlorimuron-ethyl/ metribuzin) Mfg: DuPont

Carpetweed, cocklebur, copperleaf, hophorn bean, jimsonweed, lambsquarters, mustard, pigweed, prickly sida, purslane, ragweed, smartweed, spotted spurge, sunflower, velvetleaf.

PRIMISULFURON-METHYL, BEACON, RIFLE Mfg: Ciba

Bur cucumber, shattercane, johnsongrass, quackgrass, fall panicum, annual ragweed, quack grass, ragweed, cocklebur, devils claw, nightshade, Florida beggarweed, Jerusalem artichoke, jimsonweed, kochia, ladysthumb, smartweed, pigweed, prickly sida, sicklepod, sunflower, velvetleaf, foxtails, sandbur, horseweed, horsenettle, puncture vine, wild radish, wild mustard, volunteer alfalfa, volunteer sorghum, yellow nutsedge, annual ryegrass, horseweed, horse nettle, nightshade, puncturevine, wild radish, wild mustard, volunteer sorghum.

PRODIAMINE, BARRICADE, ENDURANCE Mfg: Sandoz Agro

Crabgrass, annual bluegrass, chickweed, purslane, foxtail, shepardspurse, henbit, knotweed, goosegrass, rescuegrass, barnyardgrass, signalgrass, carpetgrass, lambsquarters, goosegrass, henbit, johnsongrass, pigweed, spurge, yellow woodsorrel.

PROMETONE, PRAMITOL   Mfg: CIBA

Annual weeds, bermudagrass, bindweed, bromegrass, goldenrod, goosegrass, Johnsongrass, oatgrass, plaintain, puncturevine, quackgrass, wild carrot.

PROMETRYNE, CAPAROL   Mfg: CIBA

Annual morningglory, barnyardgrass, carpetweed, carelessweed, cranes-bill, cheatgrass, cocklebur, coffeeweed, crabgrass, chickweed, fiddleneck, field pennycress, fleabane, Florida pusley, foxtails, galinsoga, goosegrass, groundcherry, groundsel, horseweed, jimsonweed, jungle rice, knotweed, kochia, lady's thumb, lambsquarters, little barley, malva, mayweed, mustards, nightshade, pigweed, pineappleweed, prickly sida, purslane, ragweed, rattail fescue, Russian thistle, sandbur, shepherd's-purse, signalgrass, skeletonweed, smartweed, spurred anoda, sunflower, teaweed, velvetleaf, velvetgrass, water hemp.

PRONAMIDE, KERB   Mfg: Rohm & Haas

Annual morningglory, barnyardgrass, bentgrass, bluegrass, bromes, burning nettle, canarygrass, carpetweed, cheat, chickweed, crabgrass, dodder, fall panicum, fescues, foxtail barley, foxtails, goosefoot, goosegrass, henbit, knotweed, jointed goatgrass, lambsquarters, London rocket, lovegrass, mallow, morningglory, mustards, nettles, nightshade, orchardgrass, purslane, quackgrass, red sorrel, ryegrass, shepherd's-purse, smartweed, velvetgrass, volunteer grains, volunteer tomatoes, wild oats, wild radish.

PROPACHLOR, RAMROD   Mfg: Monsanto

Annual ryegrass, barnyardgrass, carpetweed, crabgrass, fall panicum, Florida pusley, giant foxtail, goosegrass, green foxtail, groundsel, lambsquarters, pigweed, purslane, ragweed, sandbur, shepherd's-purse, smartweed, watergrass, wild buckwheat, yellow foxtail.

PROPANIL, STAM, WHAM,   Mfg: Rohm & Haas,
PROSTAR   Cedar Chem and others

Alligatorweed, barnyardgrass, Brachiaria, burhead, crabgrass, curly dock, curly indigo, foxtails, goosegrass, gulf cockspur, Hoorahgrass, jointed sedge, kochia, lambsquarters, Mexican weed, parpagrass, pigeongrass, pigweed, redstem, redweed, spearhead, spike rush, tall indigo, Texas millet, watergrass, wild buckwheat, wild mustard, wiregrass, wooly croton, Texas panicum, cocklebur, hemp seshamia, croton, signalgrass, Texas weed, paragrass, buckwheat.

PRYRAZON, PYRAMIN   Mfg: BASF

Buffalobur, chickweed, fanweed, fiddleneck, goosefoot, groundcherry, groundsel, henbit, knotweed, lambsquarters, London rocket, mayweed, mustard, nettleleaf, nightshade, pennycress, pigweed, prickly lettuce, purslane, ragweed, shepherd's-purse, smartweed, sowthistle, wild buckwheat, wild carrot.

PUCCINIA CANNLICULATA,   Mfg: Tifton Innovations
DR. BIOSEDGE

Yellow nutsedge.

PURSUIT PLUS (imazethapyr/   Mfg:American Cyanamid
pendemethalin)

Barnyardgrass, alligator weed, carpetweed, cocklebur, crabgrass, crowfootgrass, Florida pusly, foxtails, field sandbur, galinsoga, goosegrass, jimsonweed, kochia, lambsquarters, mustards, nightshade, panicums, pigweed, prickly sida, puncture vine, purslane, seedling Johnsongrass, shattercane, signalgrass, small flower morningglory, smartweed, spurges, sunflower, tall waterhemp, velvetleaf, Venice mallow, wild poinsetta, witchgrass, wooly cupgrass.

PYRIDATE, TOUGH, LENTAGRAN   Mfg: Sandoz

Palmer amaranth, spiny amaranth, Florida beggarweed, carpetweed, chickweed, cocklebur, hemp sesbania, jimsonweed, lambsquarters, nightshade, pigweed, purslane, ragweed, coffee senna, velvetleaf, bedstraw, corn chamomile, corn spurry, dead nettle, galsinoga, henbit, kochia, mayweed, morninglory, shepardspurse, yellow rocket.

QUINCLORAC, FACET   Mfg: BASF

Barnyardgrass, junglegrass, broadleaf signalgrass, crabgrass, jointvetch species, morninglory, sesbania, eclipta.

QUIZALOFOP-ETHYL, ASSURE II   Mfg: DuPont

Barnyardgrass, bermudagrass, broadleaf signalgrass, crabgrass, crowfootgrass, fall panicum, foxtails, goosegrass, itchgrass, Johnsongrass, jungle rice, quackgrass, red rice, sandbur, shattercane, sprangletop,

Texas panicum, volunteer cereals, volunteer corn, wild oats, wild proso millet, wirestem muley, witchgrass, wooly cupgrass.

RESOLVE (imazethapyr/ dicamba)   **Mfg: American Cyanamid**

Alligatorweed, spurral anoda, jerusalem artichoke, wild buckwheat, buffalo9bur, bristly starbur, carpetweed, cocklebur, jimsonweed, knotweed, kochia, lambsquarters, Venice mallow, marshelder, morniglory, mustard, nightshade, pigweed, puncturevine, purslane, Florida pusley, ragweed, barnyard sage, prickly seda, smartweed, spurge, sunflower, velvetleaf, Canada thistle, tall waterbugs, barnyardgrass, crabgrass, wooly cupgrass, foxtails, johnsongrass, wild prose millet, red rice, field sandbur, shattercane, broadleaf signalgrass, sorghum, nutsedge.

ROUT (oxyfluorfen/oryzalin)   **Mfg: Scotts**

Annual bluegrass, barnyardgrass, bittercress, chickweed, crabgrass, dandelion, fall panicum, fireweed, foxtails, goosegrass, groundsel, lambsquarters, marestail, oxalis, pearlwort, pigweed, purslane, shepherd's-purse, sow thistle, spurge.

SALUTE (metribuzin/trifluralin)   **Mfg: Miles Inc.**

Barnyardgrass, beggarweed, bluegrass, bristly starbur, browntop millet, carpetgrass, cooperleaf, crabgrass, crowfootgrass, Florida pusley, foxtails, galinsoga, goosegrass, hophorn bean, jimsonweed, Johnsongrass, jungle rice, knotweed, kochia, lambsquarters, mustard, panicums, pigweed, prickly sida, purslane, ragweed, redweed, Russian thistle, sandbur, sesbania, shepherd's-purse, signalgrass, smartweed, sprangletop, spurred anoda, stinkgrass, velvetleaf, Venice mallow.

SCEPTER O.T. (imazaquin/ acifluorfen)   **Mfg: American Cyanamid**

Cocklebur, hemp sesbania, morningglory, pigweed, smooth waterhemp, tall waterhemp.

SETHOXYDIM, POAST, VANTAGE, TORPEDO   **Mfg: BASF**

Bahiagrass, barnyardgrass, bermudagrass, crabgrass, cupgrass, foxtails, goosegrass, itchgrass, Johnsongrass, jungle rice, lovegrass, panicums, orchardgrass, quackgrass, red rice, ryegrass, signal grass, sprangletop, tall fescue, volunteer cereals, volunteer corn, wild cane, wild oats, wild proso millet, wirestem muhly, witchgrass, wooly cupgrass.

SIDURON, TUPERSAN   **Mfg: DuPont**

Crabgrass, foxtails, barnyardgrass.

SIMAZAT (simazine/atrazine)   **Mfg: Drexel**

Most annual grasses and broadleaf weeds.

SIMAZINE, PRINCEP   **Mfg: CIBA and others**

Algae (unicellular and filamentous), alyssum, amaranths, annual bluegrass, annual morningglory, annual ryegrass, barnyardgrass, Brachiaria, burclover, burdock, Canada thistle, carelessweed, carpetweed, cheatweed, chickweed, coontails, crabgrass, dogfennel, downybrome, duck salad, fall panicum, fiddleneck, filaree, fireweed, five hook brassia, Flora's paintbrush, Florida pusley, foxtail, goosegrass, groundsel, henbit, junglerice, knawel, lambsquarters, mustard, naiad, nightshade, pepperweed, pigweed, pineappleweed, plaintain, prickly lettuce, purslane, quackgrass, ragweed, rattail fescue, redmaids, Russian thistle, shepherd's-purse, shieldcress, signalgrass, silverhair grass, smartweed, Spanish needles, speedwell, submerged aquatic weeds, tansy mustard, wild oats, wiregrass, witchgrass, yellow flower pepperweed.

SNAPSHOT (isoxaben/oryzalin)   **Mfg: DowElanco**

Aster, barnyardgrass, annual bluegrass, bursage, wild celery, chickweed, clover, crabgrass, cudweed, cupgrass, fiddleneck, filaree, fleabane, yellow foxtail, groundcherry, groundsel, henbit, horseweed, jungle rice, knotweed, lambsquarters, mallow, morningglory, mustards, nightshade, wild oats, fall panicum, Virginia pepperweed, pigweed, pineappleweed, plaindain, purslane, wild radish, ragweed, London rocket, rock primrose, ryegrass, shepards purse, sibara, smartweed, sowthistle, speedwell, spurge, telegraph plant, Russian thistle, velvetleaf, witchgrass, wild carrot, dandelion, giant foxtail, galinsoga, prickly lettuce, lovegrass, marestail, prickly sida, sweet clover, yellow wood sorrel, brome grass, bur clover, hare barley, Datura, curly dock, evening primrose, fescue, jimsonweed, kochia, black medic, nettle bristly ox tongue, Scarlet pimernel, sunflower, swinecress, willow weed.

SNAPSHOT TG (trifluralin/isoxaben) Mfg: DowElanco

Aster, barnyardgrass, annual bluegrass, annual bursage, wild celery, chickweed, clover, crabgrass, cudweed, fiddleneck, filaree, fleaabane, foxtails, ground cherry, henbit, horseweed, junglerice, knotweed, lambsquarters, mallow, mustards, nightshade, wild oats, fall panicum, pepperweed, pigweed, pineapple weed, plantain, pruslane, wild radish, ragweed, London rocket, rock purslane, shepardspurse, smartweed, sibara, sowthistle, speedwell, spurge, telegraphplant, Russian thistle, velvetleaf, witchgrass, bittercress, brassbuttons, wild carrot, chamberbitter, dandelion, eclipta, galinsoga, Carolina geranium, groundsel, lovegrass, prickly lettuce, marestail, mayweed, morninglory, pennwort, pokeweed, annual ryegrass, prickly sida, red sonel, sweet clover, tansy mustard, yellow wood sorrel, hare barley, bromegrass, bur clover, datura, dog fennel, evening primrose, fescue, goosefoot, jimsonweed, knotweed, kochia, black medic, turkey mullin, nettle, bristly ox tongue, sandbur, stinkgrass, sunflower, swinecress, musk thistle, willow weed.

SODIUM CHLORATE, DEFOL Mfg: Drexel, Kerr-McGee and others

Bermudagrass, bindweed, Canada thistle, hoary cress, Johnsongrass, leafy spurge, most annual weeds, paragrass, quackgrass, Russian knapweed.

SQUADRON (imazaquin/pendimethalin) Mfg: American Cyanamid

Barnyardgrass, bristly starbur, bur cucumber, carpetweed, cocklebur, crabgrass, crowfootgrass, Florida pusley, foxtails, goosegrass, jimson-weed, Johnsongrass, kochia, lambsquarters, morningglory, mustards, nightshade, panicums, pigweed, puncturevine, purslane, ragweed, red weed, sandbur, shattercane, signalgrass, smartweed, spurge, sunflower, teaweed, velvetleaf, Venice mallow, wild poinsettia, witchgrass.

STAMPEDE CM (propanil/MCPA) Mfg: Rohm & Haas

Foxtails, wild buckwheat, kochia, lambsquarters, mustards, pigweeds.

STORM (bentazon/acifluorfen) Mfg: BASF

Annual ragweed, bristly starbur, cocklebur, crotalaria, giant ragweed, jimsonweed, lady's thumb, lambsquarters, morningglory, pigweed, prickly sida, redweed, sesbania, smartweed, spurred anoda, tall waterhemp, velvetleaf, Venice mallow, wild mustard.

SULFCARBAMIDE, ENQUIK   Mfg: Unocal

Annual fleabane, arrowhead sida, balsam apple, bracken fern, bristly starbur, bur gherkin, Carolina geranium, carpetweed, chicory, coffee senna, bedstraw, chickweed, cocklebur, lambsquarters, cudweed, evening primrose, dandelion, dock, eclipta, fiddleneck, filaree, Florida beggarweed, groundcherry, henbit, horseweed, Japanese honeysuckle, jimsonweed, knotweed, kochia lantana, London rocket, lupine, malva, milkweed, miners lettuce, morningglory, mullein, mustards, nettleleaf goosefoot, nightshade, smartweed, pigweed, pineapple weed, plaintains, poison hemhock, poison oak, prickly lettuce, prickly sida, puncturevine, red clover, shepardspurse, sicklepod, sowthistle, Spanish needles, spiny amaranth, spurge, stinging nettle, sumac, sunflower, sweet clover, swine cress, telegraph plant, tropic croton, Velvetleaf, white pellitory, wild cucumber, wild poinsettia, wild radish, wooly croton, wild oats.

SULFOMETURON-METHYL, OUST   Mfg: DuPont

Annual bluegrass, anise, bahiagrass, barnyardgrass, bouncing bet, bromes, buckhorn plaintain, bur clover, Canadian thistle, canarygrass, Carolina geranium, chickweed, clovers, curly dock, dandelion, dewberry, dogfennel, doveweed, fescues, fiddleneck, filaree, fireweed, fleabane, foxtails, foxtail barley, goldenrod, groundsel, hemlock, hemp dogbane, hoary cress, honeysuckle, horsetail, horseweed, Jerusalem artichoke, Kentucky bluegrass, kochia, kudzu, lambsquarters, little barley, mallow, mayweed, milkthistle, muskthistle, mullein, mustards, ox-eye daisy, panicums, pigweed, plantain, poison ivy, pokeweed, prickly lettuce, puncturevine, ragweed, Russian thistle, ryegrass, sowthistle, speedwell, sprangletop, bedstraw, St. John's wort, starthistle, sunflower, sweetclover, tansy mustard, tansy ragwort, turkey mussein, vetch, Virginia pepperweed, volunteer wheat, white snakeroot, wild blackberry, wild carrot, wild oats, yellow nutsedge, yellow rocket.

SURPASS 100 (acetochlor/atrazine)   Mfg: Zeneca

Nightshade, kochia, carpetweed, cocklebur, ragweed, beggarweed, Florida pusley, galensoga, giant ragweed, jimsonweed, lambsquarters, morninglory, prickly sida, purslane, pigweed, sicklepod, smartweed, tall waterbugs, velvetleaf, barnyardgrass, broadleaf signalgrass, panicums, brabgrass, crowfootgrass, sandbur, foxtail millet, goosegrass, red rice, sprangletop, johnsongrass, shattercane, wild prose millet, witchgrass, cupgrass, yellow nutsedge.

SUTAZINE (butylate/atrazine)   Mfg: Zeneca

Barnyard grass, broadleaf, signalgrass, crabgrass, fall panicum, sandbur, foxtails, goosegrass, johnsongrass, Texas panicum, volunteer sorghum, shattercane, wild oats, witchgrass, annual morningglory, carpetweed, cocklebur, lambsquarters, purslane, Florida pusley, jimsonweed, mustards, nightshade, pigweed, ragweed, sicklepod, coffeeweed, smartweed, velvetleaf, hemp dogbane, nutsedge, bermudagrass.

SYNCHRONY STS   Mfg: DuPont
(chlorimuron-ethyl/thifensulfuron/methyl)

Beggarticks, bristly starbur, bur cucumber, cocklebur, milkweed, coropea, Florida beggarweed, Florida pusley, hemp sesbania, lambsquarter, Terisalem, artichoke, jimsonweed, morninglory, mustard, pigweed, ragweed, sicklepod, smartweed, sunflower, tall waterhemp, wild poinsettia, yellow nutsedge, velvetleaf.

TEAM (benefin/trifluralin)   Mfg: DowElanco

Barnyardgrass, crabgrass, foxtails, goosegrass, poa annua.

TEBUTHIURON, GRASLAN, SPIKE   Mfg: DowElanco

Alfalfa, alkali sida, asters, barley, barnyardgrass, bedstraw, bermudagrass, black medic, bluegrass (annual), bluegrass (Kentucky), bouncing bet, bristly oxtongue, bromegrasses, broomsedge, buckhorn plantain, buffelgrass, bull sedge, burclover, buttercup, camphorweed, Canada thistle, Carolina geranium, carpetweed, catsear, cheat, chickweed, chicory, cinquetail, clovers, cocklebur, common mullein, common ragweed, common reed, crabgrass, croton, crowfootgrass, cudweed, dallisgrass, dandelion, dock, dogfennel, fall panicum, fescues, fiddleneck, filaree, fivehook bassia, fleabane, foxtail, giant ragweed, goldenrod, grape, gumweed, henbit, horseweed, itchgrass, Japanese honeysuckle, Johnsongrass, knapweed, knotweed, kochia, lambsquarters, little barley, lovegrass, lupine, mallow, milkweed, morningglory, mustard, nightshade, orchardgrass, pepperweed, pigweed, plantain, poison hemlock, poison ivy, poor joe, prickly lettuce, prickly sida, puncturevine, purslane, raspberry, reed canarygrass, rosering gaillardia, Russian knapweed, Russian thistle, ryegrass, saltbush, sandbur sedge, shepherd's-purse, smartweed, sow histle, spikeweed, spurge, staghorn, sumac, starthistle, strawberry, sunflower, swamp smartweed, telegraph plant, Texas panicum, timothy, trembling aspen, triple awngrass, trumpet creeper, vaseygrass, velvetgrass, venue looking glass, vetch, Virginia creeper, Virginia pepperweed, western ragweed, wild carrot, wild oat, wild parsnip, witchgrass, yellow wood sorrel, and most woody species.

TERBACIL, SINBAR   Mfg: DuPont

Annual bluegrass, bermudagrass, chickweed, crabgrass, crowfoot, dog fennel, fireweed, Flora's paintbrush, Florida pusley, guineagrass, henbit, horsenettle, Johnsongrass (from seed), jungle rice, knotweed, lambsquarters, mustard, natalgrass, nightshade, pigweed, panicum, purslane, ragweed, sandbur, smartweed, sheep sorrel, quackgrass, watergrass, wild geranium, yellow nutgrass.

THIOBENCARB, BOLERO, ABOLISH   Mfg: Valent

Barnyardgrass, duck salad, false pimpernel, redstem, small flower, spike rush, sprangletop, umbrella sedge, water hyssop.

TILLER (fenoxaprop-ethyl/   Mfg: Agr Evo
2,4-D/MCPA)

Annual sowthistle, field pennycress, foxtails, lambsquarters, pepperweed, pigweed, proso millet, purslane, ragweed, shepards purse, stinkweed, tumble mustard, wild mustard, wild oats, yellow rocket.

TOPSITE (imazapyr/diuron)   Mfg: American Cyanamid

Most annual grasses and broadleaf weeds.

TORNADO (fluaziflop-butyl/formesafen)   Mfg: Zeneca

Carpetweed, crotalaria, jimsonweed, ladysthumb, lambsquarters, morninglory, mustard, nightshade, pigweed, purslane, ragweed, giant ragweed, smartweed, Venice mallow, yellow rocket, barnyardgrass, broadleaf, signalgrass, crabgrass, downy brone, fall panicum, sandbur, foxtails, goosegrass, ryegrass, itchgrass, johnsongrass, shatter cane, junglerice, sorghum, Texas panicum, volunteer cereals, wild proso millet, witchgrass, wild oats, wooly cupgrass.

TRIALLATE, FARGO,   Mfg: Monsanto

Wild oats.

TRIBENURON-METHYL, EXPRESS   Mfg: DuPont

Bushy wallflower, chickweed, corn spurry, false chamomile, fiddleneck, field pennycress, flixweed, groundsel, Kochia, lambsquarters, mayweed, miners lettuce, mustards, pineappleweed, prickly lettuce, Russian thistle, treacle mustard, wild chamomile.

TRICLOPYR, GARLON, TURFLON,   Mfg: DowElanco
REMEDY, REDEEM, RELY, PATHFINDER,
GRANDSTAND

Alder, arrowwood, ash, aspen, beech, birch, blackberry, black brush, black gum, black medic, burdock, bull thistle, burdock, cactus, Canadian thistle, cascara, cherry, chicory, chinquapin, choke cherry, clanothus, clover, cottonwood, dandelion, dogwood, Douglas fir, elderberry, elm, field bindweed, goldenrod, granjino, ground ivy, guajillo, hawthorn, hazel, hickory, hornbean, huisache, lambsquarters, lespedeza, locust, maple, matchweed, modrone, mulberry, mustard, oaks, oxalis, persimmon, plantain, pine, poison oak, poplar, ragweed, salmonberry, sassafras, scotch broom, smartweed, sumac, sweetboy magnolia, sweet clover, sweet gum, sycamore, tan oak, tansy ragwort, thimbleberry, trumpet creeper, tulip poplar, twisted acacia, vetch, wild carrot, wild lettuce, wild rose, wild violet, willow, winged elm, yarrow, arlanthus, basswood, blackberry, box elder, Brazilian pepper, gallberry, guava, hercules club, manzanita, mesquite, mountain lauel, autumn olive, Russian olive, poison ivy, balsam, cedar, sweetgum, tamarack, walnut, yaupon, yellow poplar, alligatorweed, Texasweed sesbania, Northern jointvetch, redstem, morninglory.

TRIFLURALIN, TREFLAN,   Mfg: DowElanco & others
TRUST, TRILIN, TRI-4

Annual bluegrass, barnyardgrass, Brachiaria, bromegrass, carelessweed, carpetweed, cheat, chickweed, crabgrass, Florida pusley, foxtails, goosefoot, goosegrass, Johnsongrass, junglerice, knotweed, kochia, lambsquarters, pigweed, purslane, Russian thistle, sandbur, sprangletop, stinging nettle, stinkgrass, Texas panicum, wild barley, wildcane, witchweed, wooly cupgrass.

TRI-SCEPT (imazaquin/trifluralin)   Mfg: American Cyanamid

Barnyardgrass, bristly starbur, broadleaf signalgrass, bur cucumber, carpetweed, cocklebur, crabgrass, crowfootgrass, Florida pusly, foxtails, giant ragweed, goosegrass, jimsonweed, Johnsongrass seedlings, kochia, lambsquarters, morningglory, mustards, nightshade, panicums, pigweed, prickly sida, puncturevine, purslane, ragweed, redweed, sandbur, shattercane, smartweed, sunflower, velvetleaf, Venice mallow, wild poinsetta, wooly cupgrass.

TRISULFURON, AMBER   Mfg: Ciba

Mustards, field pennycress, flixweed, shepards purse, Virginia pepperweed, ragweed, sunflower, kochia, prickly lettuce, wild radish, bushy wall flower, buttercup, fiddleneck, corn gromwell, evening primrose, hairy vetch, chickweed, London rocket, marshelder, lettuce, pigweed, whitlowgrass, Jacob's-ladder, mallow, forget-me-not, Russia thistle, wild buckwheat, henbit, wild garlic, wild onion, annual ryegrass, cheat, downy brome, Canada thistle.

TRIFENSULFURON-METHYL,   Mfg: DuPont
PINNACLE

Annual knavel, annual sowthistle, buckwheat, bushy wallflower, Carolina geranium, chickweed, coast fiddleneck, common buckwheat, corn chamomile, corn spurry, curly dock, dogfennel, false chamomile, fiddleneck, field pennycress, flixweed, groundsel, knotweed, kochia, lady's thumb, lambsquarters, lentils, London rocket, mallow, malva, marshelder, mayweed, miner's lettuce, mousear chickweed, mousear cress, mustards, peas, pigweed, red maids, Russian thistle, shepards purse, smallflower buttercup, smartweed, sunflower, swinecress, tarweed, treadle mustard, tumble mustard, velvetleaf, wild buckwheat, wild chamomile, wild garlic.

TURBO (metribuzin/metolachlor)   Mfg: Miles Inc.

Barnyardgrass, beggarweed, bluegrass, bristly starbur, carpetweed, copperleaf, crabgrass, crowfoot grass, cupgrass, Florida pusley, foxtails, galinsoga, goosegrass, jimsonweed, jungle rice, knotweed, kochia, lambsquarters, mustards, nightshade, panicum, pigweed, prickly sida, purslane, ragweed, red rice, Russian thistle, sesbania, shepherd's-purse, sicklepod, signalgrass, smartweed, spotted spurge, spurred anoda, velvetleaf, Venice mallow, witchgrass.

TURFLON D (triclopyr/2,4-D)   Mfg: DowElanco

Black medic, bullthistle, burdock, buttercup, Canadian thistle, carpetweed, catnip, chamise, chickweed, chicory, cinquefoil, clovers, cocklebur, coffeeweed, cornflower, cornspeedwell, creeping beggarweed, dandelion, docks, dwarf beggarweed, field bindweed, goldenrod, ground ivy, henbit, hop clover, knawels, lambsquarters, lespedeza, little starwort, mallow, matchweed, mustards, oxalis, parsley, piert, plantain, poison ivy, poison oak, prostrate spurge, purslane, smartweed, sowthistle, speed-well, spiderwort, spotted catsear, spurweed, vetch, wild carrot, wild violet, yarrow.

TURFLON II AMINE (2,4-D/triclopyr)   Mfg: DowElanco

Black medic, carpetweed, catnip, chamise, chickweed, chicory, cinque-foil, clovers, cocklebur, coffeeweed, cornflower, beggarweed, dandelion, day flower, dock, field bindweed, burdock, buttercup, mustard, cornspeedwell, ground wig, oxalis.

TYPHOON (fluazifop-p-butyl/   Mfg: Zeneca
fomesafen)

Amaranth, spurrel anoda, carpteweed, citron, cocklebur, hophornbean copperleaf, Virginia copperleaf, crotalaria, tropic croton, volunteer cu-cumber, eclipta, jimsonweed, ladythumb, lambsquarter, Mexicanweed, morninglory, mustard, nightshade, yellow nutsedge, pigweed, purslane, wild poinsettia, Florida pusley, ragweed, giant ragweed, redweed, hemp sesbania, prickly sida, smartweed, smellmelon, spurge, bristly starbur, velvetleaf, Venice mallow, tall waterhemp, witchgrass, yellow rocket, wild oats, wild prose millet, witchgrass, cupgrass, ryegrass, itchgrass, johnsongrass, junglerice, shattercane, sorghum, Southern sandbur, Texas panicum, volunteer cereals, barnyardgrass, broadleaf signalgrass, crab-grass, downy brome, fall panicum, sandbur, foxtails, goosegrass.

VERNOLATE, VERNAM   Mfg: Drexel

Annual morningglory, balloonvine, barnyardgrass, carpetweed, cockle-bur, coffeeweed, crabgrass, fall panicum, Florida pusley, foxtails, Ger-man millet, goosegrass, Johnson grass, lambsquarters, nutgrass, pig-weed, purslane, sandbur, sicklepod, velvetleaf, wild cane.

WEEDMASTER   Mfg: Sandoz
(dicamba/2,4-D)

Beebalm, bindweed, broomweed, buckeye, buffalobear, burdock, butter-cup, chickweed, chicory, clovers, cocklebur, wooly croton, coreopsis, dandelion, dewberry, curly dock, dog fennel, elderberry, fleabane, gold-enrod, goldenweed, groundsel, henbit, honeysuckle, horsenettle, poison ivy, knapweeds, knotweed, kochie, lambsquarters, morninglory, mus-tards, nightshade, pennycress, pepperweed, persimmon, pigweed, poor joe, pivet, rabbitbush, ragweed, ragwort, redvine, sage, sneezeweed, smartweed, sorrel, sowthistle, spurge, sunglower, thistles, velvetleaf, vetch, yankeeweed, cowcockle, wild buckwheat, mallow, purslane, Russian thistle, wild carrot, tansy ragwort, Canada thistle, Carolina geranium, evening primrose, shepards purse, prickly sida, prickly lettuce, wild garlic, wild onion, asters.

XL (benefin/oryzalin)   Mfg: DowElanco

Annual bluegrass, annual ryegrass, barnyard grass, carpetweed, chickweed, crabgrass, crowfootgrass, foxtails, goosegrass, henbit, Johnson grass, knotweed, lovegrass, purslane, sandbur, wild oats.

ZINC CHLORIDE   **Mfg: Numerous**

Moss.

NOTES

FUNGICIDES

TRADE NAME CONVERSION TABLE — FUNGICIDES

Agri-Strep	*Streptomycin*
Agrimycin	*Streptomycin*
Alamo	*Propiconazole*
Aliette	*Fosetyl-Al*
Anchor	*Oxadixyl*
Apron	*Metalaxyl*
AQ-10	*Ampelomyces quioqudis*
Arasan	*Thiram*
Arbortect	*Thiabendazole*
Banner	*Propiconazol*
Banol	*Propamocarb-Hydrochloride*
Bayleton	*Triadimefon*
Baytan	*Triadimenol*
Benlate	*Benomyl*
Binab-T	*Trichoderma harzianum*
Blue-Shield	*Copper Hydroxide*
Botran	*DCNA*
Bravo	*Chlorothalonil*
Busan-30	*TCMTB*
Carbamate	*Ferbam*
Chipco-26019	*Iprodione*
Cleary 3336	*Thiophanate*
COCS	*Copper Oxychloride Sulfate*
Copper Count-N	*Copper Ammonium Complex*
COS	*Copper Oxychloride*
Cupric Zinc Sulfate Complex	*Zinc Coposil*
Curalan	*Vinclozolin*
Daconil 2787	*Chlorothalonil*
Demosan	*Chloroneb*
Dithane M-45	*Mancozeb*
Dividend	*Difenoconazole*
Domain	*Thiophanate-methyl*
Du-Ter	*Fentin Hydroxide*
Eagle	*Myclobutanil*
Echo	*Chlorothalonil*
Enable	*Fenconazole*
Ecip	*B. Subtillis*
Folicur	*Tebuconazole*
Flo-Pro	*Imazalil*
Fore	*Mancozeb*
Funginex	*Triforine*
Fungo	*Thiophanate-Methyl*
Gus 2000	*Bacillus subtilis*
Indan	*Fenconazole*

129

K-Cop	*Copper Ammonium Complex*
Koban	*Etridiazol*
Kocide	*Copper Hydroxide*
Kodiak	*S. subtilis*
Lime Sulfur	*Calcium Polysulfide*
Manzate 200	*Mancozeb*
Mertect	*Thiabendazole*
Milban	*Dodemorph-acetate*
Mycoshield	*Terramycin*
Mycostop	*Streptomyces grisevirides*
Nordox	*Copper Oxide*
Nova	*Myclobutanil*
Nusan	*TCMTB*
Nuzone	*Imazalil*
Orbit	*Propiconazol*
Ornalin	*Vinclozolin*
Penncozeb	*Mancozeb*
Phyton 27	*Copper Sulfate*
Pipron	*Piperalin*
Polyram	*Metiram*
Procure	*Triflumizole*
Prostar	*Flutolonil*
Protect	*Mancozeb*
Rally	*Myclobutanil*
Ridomil	*Metalaxyl*
Ronilan	*Vinclozolin*
Rovral	*Iprodione*
Rubigan	*Fenarimol*
Sentinal	*Cyproconazole*
Start	*Fosetyl-Al*
Stop-Scald	*Ethoxyquin*
Subdue	*Metalaxyl*
Super-Tin	*Fentin hydroxide*
Syllit	*Dodine*
Tenn-Cop	*Copper (fatty & rosin acids)*
Terraclor	*PCNB*
Terraguard	*Triflumizole*
Terraneb	*Chloroneb*
Tilt	*Propiconazol*
Topsin	*Thiophanate*
Topsin-M	*Thiophanate-Methyl*
Tri-Basic	*Basic Copper Sulfate*
Truban	*Etridiazol*
Turfcide	*PCNB*
Vapam	*Metham*
Victus	*Pseudomonas fluorescens*
Vitavax	*Carboxin*

FUNGICIDES

AG PREMISES
Binab-T
Coal tar compounds
Copper Napthenate
O-Phenylphenol

ALFALFA
Anchor
Apron
Basic copper sulfate
Copper hydroxide
Lime Sulfur
Nordox
Ridomil
Sulfur
Thiram
Top Cop

ALMONDS
Basic copper sulfate
Benlate
Bordeaux mixture
Captan
COCS
Copper Count-N
Copper hydroxide
Funginex
Gallex
Galltrol-A
Lime Sulfur
Maneb
Nordox
Ridomil
Rovral
Sulfur
Topcop
Topsin-M
Ziram

APPLES
Aliette
Banner
Basic copper carbonate

Basic copper sulfate
Bayleton
Benlate
Bordeaux Mixture
Botran
Captan
COCS
Copper Count-N
Copper hydroxide
Diphenylamine
Ferbam
Funginex
Gallex
Galltrol-A
Lime Sulfur
Mancozeb
Maneb
Mertect
Nordox
Polyram
Procure
Rally
Ridomil
Rubigan
Streptomycin
Sulfur
Syllit
Tenn-cop
Thiram
Topsin-M
Ziram

APRICOTS
Basic copper carbonate
Basic copper sulfate
Benlate
Bordeaux mixture
Botran
Bravo
Captan
COCS
Copper Count-N
Copper hydroxide

Funginex
Gallex
Galltrol-A
Nordox
Orbit
Ridomil
Ronila
Rovral
Tilt
Topcop
Topsin-M
Ziram

ASPARAGUS
Aliette
Funginex
Mancozeb
Ridomil
Sulfur
Topcop

AVOCADOS
Aliette
Basic copper sulfate
Benlate
Copper Count-N
Copper hydroxide
Nordox
Ridomil
Sulfur
Tenn-cop
Topcop

BALSAM PEAR
Bayleton

BANANAS
Baytan
Benlate
Bravo
Copper hydroxide
Imazalil
Mancozeb

Maneb
Nordox
Petroleum Oils
Sulfur
Thiram
· Tilt
Topcop
Topsin-M

BARLEY
Anchor
Apron
Basic copper sulfate
Baytan
Benlate
Busan
Captan
Copper hydroxide
Fungaflor
Mancozeb
Nordox
PCNB
Ridomil
Sulfur
Terrazole
Thiram
Tilt
Topcop
Vitavax

BEANS
Apron
Basic copper sulfate
Benlate
Botran
Bravo
Chloroneb
COCS
Copper Count-N
Copper hydroxide
Copper oxychloride
F-Stop
Maneb
Mycostop
Nordox

PCNB
Ridomil
Ridomil PC
Rovral
Sulfur
Streptomycin
Tenn-cop
Terrazole
Thiram
Topcop
Topsin-M
Vitavax

BEETS (Table)
Anchor
Apron
Basic copper sulfate
Copper oxychloride
Mycostop
Nordox
Ridomil
Tenn-cop
Thiram
Topcop

BLUEBERRIES
Benlate
Captan
Funginex
Gallex
Galltrol-A
Nordox
Ridomil
Rovral
Sulfur

BROCCOLI
Aliette
Anchor
Basic copper sulfate
Benlate
Bravo
COCS
Copper Count-N
Copper hydroxide

Maneb
Mycostop
Nordox
PCNB
Ridomil
Ridomil Bravo 81W
Rovral
Sulfur
Tenn-cop
Thiram
Topcop

BRUSSELS SPROUTS
Aliette
Anchor
Basic copper sulfate
Benlate
Bravo
COCS
Copper Count-N
Copper hydroxide
Maneb
Mycostop
Nordox
PCNB
Sulfur
Tenn-cop
Topcop

CABBAGE
Aliette
Anchor
Basic copper sulfate
Benlate
Bravo
COCS
Copper Count-N
Copper hydroxide
F-Stop
Maneb
Mycostop
Nordox
PCNB
Ridomil
Ridomil Bravo 81W

132

Tenn-cop
Thiram
Topcop

CANEBERRIES
Alliette
Basic copper sulfate
Bayleton
Benlate
Copper Count-N
Copper hydroxide
COCS
Gallex
Galltrol-A
Karathane
Lime sulfur
Nordox
Ridomil
Ronilan
Rovral
Sulfur
Tenn-cop
Topc

CARROTS
Anchor
Basic copper sulfate
Benlate
Botran
Bravo
COCS
Copper Count-N
Copper hydroxide
Mertect
Mycostop
Nordox
Ridomil
Ridomil/Copper
Rovral
Tenn-cop
Thiram
Topcop

CASSAVAS
Ridomil

CAULIFLOWER
Aliette
Anchor
Basic copper Sulfate
Benlate
Bravo
COCS
Copper Count-N
Copper hydroxide
Maneb
Mycostop
Nordox
PCNB
Ridomil
Ridomil Bravo 81W
Tenn-cop
Thiram
Topcop

CELERY
Anchor
Banner
Basic copper sulfate
Benlate
Botran
Bravo
COCS
Copper Count-N
Copper hydroxide
Mycostop
Nordox
Tenn-cop
Tilt
Topcop
Topsin-M

CHERRIES
Banner
Basic copper carbonate
Basic copper sulfate
Benlate
Bordeaux mixture
Botran
Bravo
Captan

COCS
Copper Count-N
Copper hydroxide
Ferbam
Funginex
Gallex
Galltrol-A
Lime sulfur
Nordox
Rally
Ridomil
Ronilan
Rovral
Rubigan
Sulfur
Syllit
Tenn-cop
Topcop
Topsin-M
Ziram

CHICORY
Ridomil

CHINESE CABBAGE
Benlate
Bravo
Maneb
Mycostop
PCNB
Ridomil
Topcop

CHINESE MUSTARD
Ridomil

CHINESE WAXGOURD
Bayleton

CITRON
Bayleton
Sulfur

CITRUS
Aliette

Basic copper carbonate
Basic copper sulfate
Benlate
Bordeaux mixture
Boric acid
COCS
Copper Count-N
Copper hydroxide
Copper oxychloride
Ferbam
Imazalil
Lime Sulfur
Myco-Shield
Nordox
OPP
Petroleum Oils
Ridomil
Sulfur
Tenn-cop
Topcop

CLOVER
Anchor
Apron
Lime Sulfur
Thiram

COCOA
Bravo
Copper hydroxide
Nordox

COFFEE
Bravo
Copper hydroxide
Nordox

COLLARDS
Aliette
Anchor
Basic copper sulfate
Benlate
Mycostop
Copper hydroxide
PCNB

Ridomil
Thiram

CORN
Anchor
Apron
Baytan
Benlate
Bravo
Busan
F-Stop
Mancozeb
Maneb
Tenn-cop
Terrazole
Thiram
Tilt
Topcop
Vitavax

COTTON
Anchor
Apron
Baytan
Busan
Chloroneb
Epic
Fungaflor
Mancozeb
PCNB
Ridomil
Ridomil PC
Start
Sulfur
Terrazole
Thiram
Topcop
Vitavax

COWPEAS
(Blackeyes,
southern peas)
Apron
Mycostop
Ridomil
Thiram

CRABAPPLES
Mancozeb

CRANBERRIES
Bravo
Copper Count-N
Copper hydroxide
Ferbam
Funginex
Mancozeb
Maneb
Nordox
Ridomil
Rovral
Topcop

CUCUMBERS
Aliette
Anchor
Basic copper sulfate
Bayleton
Benlate
Botran
Bravo
COCS
Copper Count-N
Copper hydroxide
Copper oxychloride
F-Stop
Funginex
Mancozeb
Maneb
Mycostop
Nordox
Reach
Ridomil
Ridomil Bravo 81W
Ridomil/Copper
Ridomil MZ-58
Ronilan
Sulfur
Tenn-cop
Thiram
Topcop
Topsin-M

134

CURRANTS
Basic copper sulfate
Benlate
COCS
Copper hydroxide
Lime Sulfur
Nordox
Rovral

DANDELIONS
Benlate

DILL
Apron
Mycostop

EGGPLANT
Anchor
Basic copper sulfate
Benlate
Captan
COCS
Copper Count-N
Copper hydroxide
Funginex
Maneb
Nordox
Ridomil
Thiram
Topcop

ENDIVE
Aliette
Botran
COC
Maneb
Ridomil
Ronilan

FENNEL
Mancozeb
Mycostop

FIGS
Mancozeb
Maneb
Sulfur

FILBERTS
Basic copper sulfate
Copper hydroxide
Copper Count-N
Gallex
Nordox

FLAX
Busan
Captan
Mancozeb

GARLIC
Benlate
Botran
Bravo
PCNB
Ridomil
Rovral
Sulfur

GENSING
Aliette
Copper hydroxide
Ridomil
Rovral

GINGER
Ridomil

GOOSEBERRIES
Basic copper sulfate
Copper hydroxide
Lime Sulfur
Nordox
Sulfur

GOURDS
Anchor
Bayleton

GRAPES
AQ-10
Basic copper sulfate
Bayleton
Benlate
Bordeaux mixture
Botran
Captan
COCS
Copper Count-N
Copper hydroxide
Ferbam
Gallex
Kocide 404S
Lime Sulfur
Mancozeb
Maneb
Nordox
Procure
Rally
Ridomil/Copper
Ridomil MZ58
Ronilan
Rovral
Rubigan
Sulfur
Tenn-cop
Topcop

GRASS
Anchor
Apron
Benlate
Bravo
Reach
Ridomil
Sulfur
Tilt

HOPS
Aliette
Basic copper sulfate
COCS
Copper hydroxide
Nordox

Ridomil
Topcop

HORSERADISH
Mycostop
Ridomil

KALE
Aliette
Anchor
Benlate
Maneb
Mycostop
PCNB
Ridomil
Thiram

KIWI
Botran
Copper hydroxide
Galltrol
Ronilan
Rovral

KOHLRABI
Anchor
Basic copper sulfate
Benlate
Maneb
Mycostop
Ridomil

KUMQUAT
Sulfur

LEEKS
Mycostop
Ridomil

LETTUCE
Aliette
Anchor
Basic copper sulfate
Botran
Captan

COCS
Copper hydroxide
Maneb
Mycostop
Nordox
Ridomil
Ronilan
Rovral
Tenn-cop
Thiram
Topcop

LONGAN
Copper compounds

LUPINE
Anchor
Apron
Ridomil

LYCHEE
Copper compounds

MACADAMIA NUTS
Benlate

MANGOES
Basic copper sulfate
Benlate
Captan
Copper Count-N
Copper hydroxide
Nordox
Sulfur
Tenn-cop
Topcop

MELONS
Aliette
Anchor
Basic copper sulfate
Bayleton
Benlate
Bravo
COCS
Copper Count-N

Copper hydroxide
Copper oxychloride
Funginex
Mancozeb
Maneb
Mycostop
Nordox
Reach
Ridomil
Ridomil Bravo 81W
Ridomil/Copper
Ridomil MZ-58
Sulfur
Tenn-cop
Thiram
Topcop
Topsin-M

MILLET
Anchor
Apron

MUSHROOMS
Anchor
Victus
Mertect

MUSTARD
Aliette
Anchor
Basic copper sulfate
PCNB
Ridomil
Rovral

NECTARINES
Banner
Basic copper carbonate
Basic copper sulfate
Benlate
Bordeaux mixture
Botran
Bravo
Captan
COCS

Copper Count-N
Copper hydroxide
Funginex
Gallex
Galltrol-A
Lime Sulfur
Nordox
Orbit
Rally
Ronilan
Rovral
Sulfur
Tenn-cop
Topcop
Topsin-M
Ziram

OATS
Anchor
Apron
Baytan
Benlate
Busan
Copper hydroxide
Mancozeb
Ridomil
Vitavax

OKRA
Apron
Thiram

OLIVES
Basic copper carbonate
Basic copper sulfate
Bordeaux mixture
Copper Count-N
Copper hydroxide
Gallex
Nordox
Topcop

ONIONS
Basic copper sulfate
Botran

Bravo
Captan
COCS
Copper hydroxide
Copper Count-N
Mancozeb
Maneb
Mycostop
Nordox
Pro-Gro
Ridomil
Ridomil Bravo 81W
Ridomil MZ-58
Ronilan
Rovral
Sulfur
Tenn-cop
Thiram
Topcop
Topsin-M

ORNAMENTALS
Alamo
Aliette
Banner
Banol
Banrot
Basic copper sulfate
Bayleton
Bordeaux mixture
Botran
Bravo
Busan-72
Captan
Chipco 26019
Cleary-3336
COCS
Copper hydroxide
Daconil 2787
Eagle
Ferbam
Fore
Funginex
Gallex
Galltrol-A

Lime Sulfur
Mancozeb
Maneb
Mertect
Mycostop
Ornalin
PCNB
Phyton-27
Pipron
Plantvax
Polyram
Reach
Ridomil
Rubigan
Subdue
Sulfur
Tenn-cop
Terraguard
Terramycin
Terrazole
Thiram
Triforine
Turban
Turfcide
Vapam
Ziram
Zyban

PAPAYAS
Basic copper sulfate
Benlate
Bravo
Mancozeb
Maneb
Mertect
Ridomil
Topcop

PARSNIPS
Anchor
Bravo
Mycostop
Ridomil

PASSION FRUIT
Bravo

PEACHES
Banner
Basic copper carbonate
Basic copper sulfate
Benlate
Bordeaux mixture
Botran
Bravo
Captan
COCS
Copper Count-N
Copper hydroxide
Copper oxychloride
Ferbam
Funginex
Gallex
Galltrol-A
Lime Sulfur
Myco-Shield
Nordox
Orbit
Rally
Ridomil
Ronilan
Rovral
Sulfur
Syllit
Tenn-cop
Terramycin
Thiram
Topcop
Topsin-M
Ziram

PEANUTS
Basic copper sulfate
Benlate
Bordeaux mixture
Bravo
COCS
Copper Count-N
Copper hydroxide

F-Stop
Folicur
Mancozeb
Nordox
PCNB
Ridomil
Ridomil PC
Rovral
Tenn-cop
Terrazole
Tilt
Topcop
Topsin-M
Vitatax

PEARS
Aliette
Basic copper sulfate
Bayleton
Benlate
Bordeaux mixture
Captan
COCS
Copper Count-N
Copper hydroxide
Ethoxyquin
Ferbam
Gallex
Galltrol-A
Lime Sulfur
Mancozeb
Mertect
Myco-Shield
Nordox
Opp
Procure
Rubigan
Streptomycin
Sulfur
Syllit
Terramycin
Topcop
Ziram

PEAS
Anchor
Apron
Apron-T69
Basic copper sulfate
Copper hydroxide
Mycostop
PCNB
Ridomil
Sulfur
Terrazole
Topcop

PECANS
Banner
Basic copper sulfate
Benlate
Du-Ter
Enable
Gallex
Galltrol-A
Orbit
Rubigan
Sulfur
Syllit
Tenn-cop
Topcop
Topsin-M
Ziram

PEPPERS
Anchor
Basic copper sulfate
Benlate
Captan
COCS
Copper Count-N
Copper hydroxide
Funginex
Maneb
Mycostop
Nordox
PCNB
Ridomil
Ridomil/Copper

Ronilan
Tenn-cop
Thiram
Topcop

PERSIMMON
Galltrol

PINEAPPLE
Aliette
Bayleton
Benlate
Mancozeb
Ridomil
Topsin-M

PISTACHIOS
Benlate

PLUMS
Banner
Basic copper sulfate
Benlate
Bordeaux mixture
Botran
Bravo
Captan
COCS
Funginex
Gallex
Galltrol-A
Kocide
Lime Sulfur
Nordox
Orbit
Ridomil
Rovral
Sulfur
Topcop
Topsin-M

POTATOES
Basic copper sulfate
Bordeaux mixture
Botran

Bravo
Bravo ZN
COCS
Copper Count-N
Copper hydroxide
Copper oxychloride
Mancozeb
Maneb
Mertect
Nordox
Polyram
Ridomil
Ridomil Bravo 81W
Ridomil/Copper
Ridomil MZ-58
Rovral
Sclerban
Sulfur
Super Tin
Tenn-cop
Topcop
Topsin-M

PRUNES
Basic copper sulfate
Benlate
Bordeaux mixture
Botran
Bravo
Captan
COCS
Funginex
Gallex
Galltrol-A
Kocide
Lime Sulfur
Nordox
Orbit
Ridomil
Rovral
Sulfur
Topcop
Topsin-M

PUMPKINS
Aliette
Anchor
Basic copper sulfate
Bayleton
Benlate
Bravo
Copper hydroxide
Maneb
Mycostop
Nordox
Ridomil
Ridomil/Copper
Tenn-cop
Topcop
Topsin-M

QUINCE
Aliette
Lime Sulfur
Mancozeb
Nordox

RADISH
Anchor
Basic copper sulfate
Copper oxychloride
Mycostop
Ridomil
Ridomil/Copper
Thiram

RAPE
Anchor
Benlate
Ridomil

RHUBARB
Botran

RICE
Apron
Benlate
Busan
Kocide

Rovral
Terrazole
Tilt
Topcop
Vitavax

RUTABAGAS
Anchor
Mycostop
Ridomil
Sulfur

RYE
Anchor
Apron
Baytan
Benlane
Mancozeb
Ridomil
Tilt

SAFFLOWER
Busan-72
Terrazole
Vitavax

SALSIFY
Mycostop
Ridomil

SESAME
Busan
Thiram

SORGHUM (Milo)
Anchor
Apron
Baytan
Busan-72
F-Stop
Ridomil
Terrazole
Thiram
Vitavax

SOYBEANS
Anchor
Apron
Benlate
Bravo
Chloroneb
F-Stop
K-Cop
Ridomil
Terrazole
Thiram
Topcop
Topsin-M
Vitavax

SPINACH
Anchor
Basic copper sulfate
Captan
COCS
Copper Count-N
Copper hydroxide
Mycostop
Nordox
Ridomil
Ridomil/Copper
Thiram
Zinc Coposil

SQUASH
Aliette
Anchor
Basic copper sulfate
Bayleton
Benlate
Bravo
COCS
Copper Count-N
Copper hydroxide
Mancozeb
Maneb
Mycostop
Nordox
Reach
Ridomil

Ridomil Bravo 81W
Ridomil/Copper
Ridomil MZ-58
Sulfur
Tenn-cop
Thiram
Topcop
Topsin-M

STRAWBERRIES
Aliette
Basic copper sulfate
Benlate
Bordeaux mixture
Captan
COCS
Copper Count-N
Copper hydroxide
Funginex
Nordox
Ridomil
Ronilan
Rovral
Sulfur
Syllit
Topcop
Topsin-M

SUGAR BEETS
Anchor
Apron
Basic copper sulfate
Bayleton
Benlate
Busan-72
Chloroneb
COCS
Copper Count-N
Copper hydroxide
Copper oxychloride
F-Stop
Mancozeb
Maneb
Mertect
Nordox

Ridomil
Ridomil MZ58
Sulfur
Super Tin
Tenn-cop
Terrazole
Topcop
Topsin-M

SUGARCANE
Tilt

SUNFLOWERS
Anchor
Apron
Captan

SWEET POTATOES
Benlate
Botran
Mertect
Ridomil
Sclerban

SWISS CHARD
Aliette
Copper oxychloride
Ridomil
Thiram

TARO
Captan

TOBACCO
COCS
Ferbam
Nordox
Ridomil

TOMATOES
Aliette
Anchor
Basic copper sulfate
Benlate
Bordeaux mixture

Botran
Bravo
Bravo C/M
COCS
Copper Count-N
Copper oxychloride
F-Stop
Mancozeb
Maneb
Mycostop
Nordox
PCNB
Ridomil
Ridomil Bravo 81W
Ridomil/Copper
Ridomil MZ-58
Ronilan
Sulfur
Tenn-cop
Topcop

TREFOIL
Anchor
Apron

TURF
Aliette
Anchor
Apron
Banner
Banol
Bayleton
Captan
Chipco 26019
Chloroneb
Cleary 3336
Daconil 2787
Fungo
Koban
Kromad
Mancozeb
Maneb
Pace
PCNB

Prostar
Rubigan
Sentinel
Subdue
Terraneb
Thiram
Topsin-M
Turfcide
Vapam
Vorlex

TURNIPS
Anchor
Basic copper sulfate
Mycostop
Ridomil
Sulfur

VETCH
Anchor
Apron

WALNUTS
Banner
Basic copper sulfate
Basic copper carbonate
Bordeaux mixture
COCS
Copper Count-N
Copper hydroxide
Gallex
Galltrol-A
Nordox
Ridomil
Tenn-cop
Topcop

WATERCRESS
Benlate

WATERMELONS
Anchor
Aliette
Basic copper sulfate
Bayleton

Benlate
Bravo
Copper Count-N
Copper hydroxide
Funginex
Mancozeb
Maneb
Nordox
Reach
Ridomil
Ridomil/Copper
Sulfur
Tenn-cop
Topcop

WHEAT
Anchor
Apron
Basic copper sulfate
Bayleton
Baytan
Benlate
Busan
Copper hydroxide
Dividend
Fungaflor
Mancozeb
Nordox
Ridomil
Sulfur
Terrazole

WILD RICE
Tilt

FUNGICIDES

AMPELOMYCES QUISQUALIS,
AQ-10   **Mfg: Ecogen**

Powdery mildew.

BACILLUS SUBTILIS, GUS 2000, EPIC   **Mfg: Gustafson**

Seedling diseases.

BANROT (etridizol/thiophanate-   **Mfg: Scotts**
methyl)

Damp off, Fusarium, Phytophtora, Pythium, Rhizoctonia, rootrot, stem
rot, Thielaviopis, water mold.

BASIC COPPER CARBONATE   **Mfg: Pace Intl.**

Blossom blight, brown rot, fire blight, leaf curl, peach blight, peacock spot,
Septoria, shot hole, walnut blight.

BASIC COPPER SULFATE,   **Mfg: Griffin Co.,**
TRI-BASIC   **Phelps Dodge, Agtrol,**
                                             Monterey and others

Alternaria leaf blight, angular leaf spot, Anthracnose, apple blotch, apple
scab, bacterial blight, bacterial canker, bacterial leaf spot, bacterial wilt,
bitter rot, black rot blight, blossom rot, blue mold, brown rot, brown rot
blossom blight, Cercospora light, Cercospora leaf spot, citrus scab,
downy mildew, early blight, fire blight, greasy spot, gummy stem blight,
late blight, leaf blight, leaf curl, leaf mold, leaf spot, melanose, nail head
rust, peach blight, peach leaf curl, peacock spot, Phomopsis, pink pitting,
powdery mildew, purple blotch, red alga, scab, Septoria leaf spot,
Septoria rot, shot hole, smut, Stemphylium leaf spot, walnut blight, white
rust, yellow rust.

BENOMYL, BENLATE   **Mfg: DuPont**

Anthracnose, basal rot, bitter rot, black rot, black spot, blackleg, blossom
blight, blue mold, Botrytis gray mold, brown patck, brown rot, bunch rot
bitter rot, Cercospora foot rot, Cercospora leaf spot, cherry leaf spot,
Cladosporium leaf mold, crown rot, Diaporthe pod, Dyslodia tip blight,
dollar spot, downy spot, dry bubble, Dutch elm disease, early blight,

Eutypa, fly speck, foot rot, Fusarium patch, Fusarium wilt, Gleosporium, glume blotch, gray mold, greasy spot, green mold, gummy stem blight, late blight, leaf spot, mummy berry, peach scab, Penicillium rots, Phoma leaf spot, pineapple butt rot, pineapple disease, powdery mildew, rice blast, rusts, scab, Sclerotinia blight, Sclerotinia white mold, sooty blotch, Septoria brown spot, Septoria leaf spot, stem blight, stem rot, stem-end rot, strawberry foot rot, stripe smut, target spot, Thielaviopsis, tip blight, white blight, white rot.

BORDEAUX MIXTURE   Mfg: Numerous

Angular leaf spot, Anthracnose, bacterial blast, bacterial blight, bacterial canker, bacterial wilt, bitter rot, black pit, black rot, blossom blight, blotch, brown rot, cane blight, canker gall, Cercospora leaf spot, crown gall, downy mildew, early blight, fire blight, frogeye leaf spot, green rot, gummosis, late blight, late scab, leaf blight, leaf mold, leaf spot, Melanrose, peach blight, peach leaf curl, peacock spot, Phytophthora, rust, scab, Septoria, shot hole, walnut blight, white mold, wild fire.

BRAVO C/M (chlorothalonil/   Mfg: ISK Biosciences
copper/maneb)

Bacterial speck, bacterial spot, early and late blight, target spot, alternaria, leaf spot anthracnose, gray leaf mold, gray leaf spot, gray mold, Septoni leaf spot.

BRAVO-ZN (chlorothalonil/   Mfg: ISK Biosciences
zinc oxide)

Early blight, late blight, Botrylis vine rot.

CALCIUM POLYSULFIDE,   Mfg: Miller Chemical
LIME SULFUR   and others

Anthracnose, aphids, apple scab, blister mite, blotch, boxwood canker, Brayobia mite, brown rot, bud mite, cane blight, canker, Coryneum blight, fruit tree leaf roller, mange, maple gall, mealybug, leaf curl, pear leaf, powdery mildew, rust scab, scales, shot hole, spider mite.

CAPTAN   Mfg: Microflo/Zeneca

Alternaria, angular leaf spot, Anthracnose, bitter rot, black pox, black rot, blotch, Botrosphaeria (white rot), Botrytis rot, Brooks fruit spot, brown patch, brown rot, bullseye rot, bunch rot, Cercospora spot, copper spot, corn rot and decay, frogeye, fruit rots, Fusarium, Gleosporium, gray leaf spot, heart rot, Helminthosporium leaf blight, jacket rot, late blight, leaf spot, Melanose, melting out, mummy berry, Penicillium, petal blight, Phoma rot, Phomopsis blight, pink rot, purple blotch, Rhizopus, root rots, rust, scab, seedling blights, Septoria leaf spot, shot hole, soot blotch, spur blight, storage rots, transit rots, tuber rot, twig and blossom blight, white mold.

CARBOXIN, VITAVAX   Mfg: Uniroyal

Cereal smuts, common bunt, covered smut, damping-off, flag smuts, loose smut, Phythium, Rhizoctonia, seedling disease, seed rot, soreshin, stripe, white mold.

CHLORONEB, DEMOSAN,   Mfg: Kincaid Mfg.
TERRANEB

Brown patch, damping-off, Phythium spp., Rhizoctonia spp., seedling blight, soreshin, Sclerotium spp.

CHLOROTHALONIL, BRAVO,   Mfg: ISK Biosciences
DACONIL 2787, EXOTHERM, ECHO   and others

Alternaria, Anthracnose, Ascochyta blight, Ascochyta ray blight, basal stalk rot, black mold, black spot, blight, blossom blight, Botrytis blight, Botrytis gray mold, Botrytis leaf spot, Botrytis vine rot, brown patch, brown rot blossom blight, brown spot, cedar apple rust, Cercospora leaf spot, cherry leaf spot, copper spot, Curvularia leaf spot, Cylindrocladrim leaf spot, Dactylaria leaf spot, dollar spot, downy mildew, early and late blight, fading out, frogeye leaf spot, fruit rot, fusarium, gray leaf spot, gray leaf mold, gray mold, gray snow mold, going out, gummy stem blight, Helminthosporium, lacy scab, leaf spot, iris leaf spot, leaf blight, leaf curl, leaf smut, Lophodermium leaf and twig blight, melting out, narrow brown leaf spot, needlecast, Ovulinia flower blight, pink rot, Phoma spp., Phythium spp., Phytophthora blight, pod and stem blight, powdery mildew, purple blotch, purple seed stain, red thread, Rhizoctonia blight, Rhizoctonia brown patch, Rhizoctonia fruit rot, Rhizoctonia spp., rice blast, ring spot, russet scab, rust, scab, scleroderris canker, Sclerotinia dollar spot, scab, Septoria leaf spot, sheeth blight, shot hole, leaf spot, Sirdcocus tip blight, Sirrhia brown spot, skin rot, Sphaeropsis leaf spot, stem rust, Swiss needle cast, tan leaf spot, target spot, tip blight, tube rot, web blotch, Helminthosporn brown blight, shothole flower spot, leaf spot, leaf blight, leaf blotch, flower blight, stem canker, scab, swiss needle cast, twig blight, foliar spot, red thread, algae scum.

COPPER (fatty & rosin acids), TENN-COP

Mfg: Miller Chemical

Leaf spots, blossom brown rot, scorch, peacock spot, anthracnose, downy mildew, Cercospora leaf spot, needle blight, Deplodia tip blight, yellow rust, cane spot, Phytophthora blight, zonate leaf spot, fireblight, bacterial canker, fruit spots, melanose, red algae, walnut blight, bacterial blights, Alternaria blight, early blight, bacterial spot, late blight, Septoria leaf spot, bacterial speck, Southern leaf blight, brown spot, halo spot, web blotch, powdery mildew.

COPPER AMMONIUM COMPLEX, COPPER COUNT-N

Mfg: Mineral Research & Dev Corp.

Alternaria, angular leaf spot, Anthracnose, bacterial blight, bacterial speck, bacterial spot, blight, blossom rot, brown rot, Cercospora leaf spot, downy mildew, early and late blight, greasy spot, halo blight, leaf scorch, leaf spot, Melanose, powdery mildew, scab, walnut blight.

COPPER HYDROXIDE, BLUE-SHIELD, KOCIDE

Mfg: Griffin, MicroFlo, Agtrol and others

Algae, Alternaria blight, angular leaf spot, Anthracnose, bacterial blast, bacterial blight, bacterial leaf blight, bacterial leaf streak, bacterial spot, ball moss, Botrytis blight, brown rot, carrot blight, Cercospora leaf spot, collar rot, common blight, common bunt, Cornyeum blight, crown rot, dead bud, downy mildew, early blight, European canker, fire blight, greasy spot, halo blight, Helminthosporium spot blotch, iron spot, late blight, leaf blotch, leaf curl, leaf rust, leaf spot, Melanose, peacock spot, Phomopsis, pink disease, pink pitting, Pseudomonos leaf spot, purple blotch, scab, seed rot, Septoria leaf blotch, shot hole, Sigatoka, Votutella leaf blight, walnut blight, water mold , Xanthomonas leaf spot.

COPPER OXIDE, NORDOX

Mfg: Nordox, Monterey Chemical

Alternaria leaf spot, angular leaf spot, Anthracnose, bacterial blast, bacterial blight, bacterial canker, bacterial speck, bacterial spot, bacterial wilt, berry spot, black leaf spot, black pod rot, black rot, blister blight, blue mold, brown rot, brown rot blossom blight, cane spot, Cercospora leaf spot, damping off, dead bud, downy mildew, early blight, fire blight, frogeye disease, fruit rot, gray leaf spot, greasy spot, gummy stem blight, halo blight, Helminthosporium spot blotch, iron spot, late blight, leaf rust, leaf spot, melanose, peacock spot, pink disease, powdery mildew, purple blotch, scab, septoria leaf blotch, septoria leaf spot, shot hole, sigatoka, walnut blight, white rust, wild fire, yellow rust.

COPPER OXYCHLORIDE, COS   Mfg: MicroFlo

Anthracnose, bacterial spot, cercospora leaf spot, damp-off, downy mildew, early blight, greasy spot, late blight, leaf curl, leaf spots, melanose, scab, Septoria.

COPPER OXYCHLORIDE SULFATE,   Mfg: Agtrol
COCS

Angular leaf spot, Anthracnose, bacterial spot, black rot, brown rot, bud blight, Cercospora leaf spot, cherry leaf spot, Coryneum blight, damping-off, downy mildew, early blight, fire blight, greasy spot, late blight, leaf blight, Macrosporum leaf spot, Melanose, peach blight, peach leaf curl, pod spot, powdery mildew, scab, Septoria leaf spot, shot hole, tar leaf spot, twig blight, walnut blight, yellow rust.

COPPER SULFATE   Mfg: Source Technology
PENTAHYDRATE, PHYTON 27

Botrytis, blackspot, cylindrocidium, Rhizoctoria, soft rot, powdery mildew, Pseudomonus, Xanthomonas, bacterial diseases, Dutch elm disease, shade tree canker, oak welt, fireblight, apple scab.

CYPROCONAZOLE, SENTINEL   Mfg: Sandoz

Brown patch, summer patch, anthracnose copper spot, gray leaf spot, gray snow mold, pink snow mold, dollar spot, necrotic ring spot, powdery mildew, red thread, rust, Southern blight, stripe smut.

DCNA, BOTRAN   Mfg: Gowan

Brown rot, Rhizopus rot, fruit decay, bunch rot, Botrytis rot, storage rot, blossom blight, white mold, gray mold, leaf rot, stem canker, pink rot, western fruit rot, botrytis blight, scurf, sclerotina blight, dry rot.

DIFENOCONAZOLE, DIVIDEND   Mfg: Ciba

Smut, bunt, Septoria seedling blight, Fusarium seed scab, general seed rots, powdery mildew, rust, Septoria leaf blotch, take-all, common foot rot, Fusarium root and crown rot.

DODINE, SYLLIT   **Mfg: Platte Chemical Co.**

Scab, liver spot, brown leafspot, leaf blotch, downy mildew, leaf curl, blossom brown rot, bacterial spot.

ETHOXYQUIN, STOP-SCALD   **Mfg: Elf-Atochem**

Scald.

ETRIDIAZOL, TERRAZOLE,   **Mfg: Uniroyal**
TRUBAN, KOBAN

Damping-off, Fusarium, Phytophthora, Phythium, stem rot, Thielaviopsis, stem rot.

FENARIMOL, RUBIGAN   **Mfg: DowElanco**

Anthracnose, copper spot, crown rot, dollar spot, Fusarium blight, Fusarium leaf spot, large brown patch, necrotic ring spot, pink and gray snow mold, poa annua, powdery mildew, red thread, scab, rust, stripe smut, summer patch.

FENCONAZOLE, ENABLE,   **Mfg: Rohm & Haas**
INDAR

Zonate leaf spot, brown rot, leaf spot, peach scab.

FENTIN HYDROXIDE, SUPER-TIN   **Mfg: Griffin**

Alternaria blight, brown leafspot, brown spot, Cercospora leaf spot, downy spot, early blight, late blight, leaf blotch, liver spot, pecan scab, powdery mildew, sooty mold.

FERBAM, CARBAMATE   **Mfg: FMC**

Scab, cedar rust, blotch, brown rot, leaf spot, lophodermium twig blight, fruit rots, black rot, blue mold, rusts, petal blight, stem rot, septoria leaf spot, alternaria blight, Botrytis blights, amthracnose.

FLUTOLANIL, PROSTAR   **Mfg: Agr Evo**

Brown patch, red thread, yellow patch, Southern blight, fairy ring, gray snow mold.

FOSETYL-AL, ALIETTE, START   Mfg: Rhone Poulenc

Brown rot, crown rot, heart rot, Phytophthora foot rot, Phytophthora root rot, Pythium spp., spear slime, pruning wound canker, Xanthomonas spp.

GALLEX   Mfg: Ag Bio Chem., Inc.

Burr knot, crown gall, olive knot, rust galls.

GALLTROL-A   Mfg: Ag Bio Chem., Inc.

Prevents crown gall.

IMAZALIL, FUNGAFLOR,   Mfg: Jannsen/Wilbur Ellis/
FLO-PRO, NUZONE   Gustafson

Alternaria citri, barley leaf stripe, blue mold, common root rot, Diplodia rot, Fusarium spp., green mold, Pennicillum spp., Phomophis stem-end rot, Thielaviopsis.

IPRODIONE, CHIPCO-26019,   Mfg: Rhone Poulenc
ROVRAL

Aerial web blight, Alternaria, black leg, Botrytis fruit rot, Botrytis leaf blight, Botrytis neck rot, black crown rot, blossom blight, cherry leaf spot, early blight, Monilinia brown rot, bottom rot, brown patch, bunch rot, Cyliadrocladrim root rot, dollar spot, early blight, Fusarium corn rot, Fusarium blight, Fusarium patch, gray mold, gray snow mold, Helminthosporium leaf spot, ink spot, lettuce drop, melting out, onion leaf blight, phomopsis, pink snow mold, purple blotch, purple leaf spot, red thread, Rhizoctonia spp., Sclerotinia, sheath blight, shot hole, soft rot, stem end rot, tulip fire, white mold, white rot, necrotic ring spot, leaf scorch.

KOCIDE 404S   Mfg: Griffin
(copper hydroxide/sulfur)

Blackrot, downy mildew, powdery mildew.

KODIAK H.B (B. subtillis)   Mfg: Gustafson

Protects against soil diseases.

MANCOZEB, DITHANE, FORE, PENNCOZEB, MANZATE 200, PROTECT

Mfg: Rohm & Haas, DuPont & Elf-Atochem

Achlia, algae, Alternaria, Anthracnose, bacterial spot, bacterial speck, bitter rot, black mold, black rot, black spot, blue mold, Bortytis blight, Botrytis leaf blight, brown rot, bunch rot, bunt, cedar apple rust, Cercospora leaf spot, copper spot, covered kernal smut, covered smut, crown rot, Curvularia leaf spot, damp-off, dead-arm, dollar spot, downy mildew, early blight, false loose smut, flyspeck, frogeye leaf spot, fruit rot, Fusarium blight, Fusarium seed decay, Fusarium snow mold, glume blotch, gray leaf mold, gray leaf spot, gummy stem blight, heart rot, Helminthosporium leaf blight, Helminthosporium melting out, Herpobasidium blight, late blight, leaf blight, leaf blotch, leaf mold, leaf spot, Lophodermium fungi, Monochaetia canker, neck rot, pecan scab, petal blight, Phytophthora, purple blotch, purple spot, Phythium blight, red thread, Rhizoctonia, brown patch, rust, scab seedling blight, seedling rot, Septoria leaf spot, Sigatoka, slime mold, sooty blotch, spodixrot, Swiss needle cast, tan spot, Taphrina leaf blister, Volutella blights.

MANEB

Mfg: Griffin, Elf-Atochem

Alternaria leaf spot, angular leaf spot, Anthracnose, bitter rot, black rot, black spot, blast, blossom blight, blue mold, Botrytis leaf blight, brown patch, brown rot, bullseye rot, Cercospora early blight, Cercospora leaf spot, Curvalaria, damping-off, dollar spot, downy mildew, early blight, flyspeck, fruit rot, gray leaf spot, green rot, gummy stem blight, Helminthosporium leaf blight, late blight, leaf spot, melting out, peach leaf curl, purple blotch, Phythium fruit rot, rust, scab, Septoria late blight, Septoria leaf spot, shot hole, Sigatoka disease, soreshin, Stemphylium, stem rust, stripe rust, twig light, white rot, white rust.

METALAXYL, APRON, RIDOMIL, SUBDUE

Mfg: CIBA and Gustafson

Blue mold, black shank, cavity spot, collar rot, crown rot, damping-off, downy mildew, early and late blight, Phytophthora blight, Phytophthora spp., Pythium spp., pod rot, root dieback, root rot, seed rot, spray rot, water mold, white rust, yellow tuft, brown patch, dollar spot.

METIRAM, POLYRAM

Mfg: Platte Chemical Co.

Early blight, late blight, black spot, scab, rusts, leaf spot.

MYCLOBUTANIL, RALLY,   **Mfg: Rohm & Haas**
NOVA, EAGLE

Blackrot, powdery mildew, scab, rust, leaf spot.

OPP, DOWCIDE   **Mfg: Dow and Elf-Atochem**

Post harvest decay organisms.

OXADIXYL, ANCHOR   **Mfg: Sandoz/Gustafson**

Phytophtora, Pythium spp.

OXYCARBOXIN, PLANTVAX   **Mfg: Uniroyal**

Rust.

PACE (metalaxyl/mancozeb)   **Mfg: Ciba**

Downy mildew, pythium blight, yellow tuft, brown patch, dollar spot.

PCNB, TERRACLOR, TURFCIDE   **Mfg: Uniroyal and Amvac**

Botrytis storage rot, bottom rot, brown patch, brown rot, bulb dry rot, bulb rot, bunt, camelia flower blight, club root, common smut, crown rot, damping-off, leaf drop, leaf spot, magnolia leaf spot, neck rot, petal blight, Rhizoctonia, root rot, Sclerotinia, snow mold, Southern blight, stem rot, white mold, white rot, wire stem, flower blight, petal blight, black rot, dry rot, melting out.

PIPERALIN, PIPRON   **Mfg:Sepro**

Powdery mildew.

PRO GRO (thiram/carboxim)   **Mfg: Uniroyal**

Onion smut.

PROPAMOCARB-   **Mfg: Agr Evo**
HYDROCHLORIDE, BANOL

Cottony blight, damping-off, grease spot, Pythium blight, Pythium spp., Phytophtora spp., root-rot, rusts.

PROPICONAZOL, BANNER, TILT, ORBIT, ALAMO
Mfg: CIBA

Antharacnose, brown leaf spot, brown patch, dollar spot, downy spot, foot rot, fungal leaf scorch, glume blotch, Helminthosporium leaf blight, leaf smut, liver spot, red blotch, pecan scab, pineapple disease, powdery mildew, red thread, rusts, scald, Selenophoma stem eyespot, Septoria leaf blight, sheath blight, stripe smut, tan spot, vein spot, Zonate leaf spot, needle rust, tip blight, gray leaf spot, pink snow mold, gray snowmold, apple scab, leaf spot, ray blight, blackspot, blossom blight, blossom blight leafspot, greasy spot, downy spot, leaf search, blue mold, leaf spots, dollar spot, brown rot, blossom blight, fruit rot, zoysia patch, early and late blight, leaf spots, leaf blight, gray leafspot eye spot.

PSEUFOMONAS FLUORESCENS, VICTUS
Mfg: Mauri Labs

Bacterial blotch.

REACH (chlorothalonil/triadimefon)
Mfg: ISK Biosciences

Powdery mildew, stem rust, leaf rust, stripe rust, Septoria leaf spot, glume blotch, Bipolaris, leaf spot, Mastigosporium eye spot, Dreschslera leafspot, stem sall rust, fusiform rust, needle casts, scleroderris canker, brown spot, tip blight.

RIDOMIL BRAVO 81W (metalaxyl/chlorothalonil)
Mfg: CIBA

Alternaria leaf spot, Antharacnose, Botrytis, Cercospora leaf spot, downy mildew, early blight, fruit rot, gray leaf mold, gray leaf spot, gummy stem blight, late blight, purple blotch, Septoria leaf spot, storage rot, vine rot.

RIDOMIL/COPPER 70W (metalaxyl/copper hydroxide)
Mfg: Ciba

Pythium spp., Phtophthora spp., late blight, taber rot, pink rot, pythium leak, cavity spot, white rust, downy mildew, bacterial speck, bacterial spot.

RIDOMIL MZ-58 (metalaxyl/mancozeb)
Mfg: CIBA

Downy mildew, late blight, Pythium leak, tuber rot.

RIDOMIL PC (metalaxyl/PCNB)　　　　　**Mfg: CIBA**

Damp-off, pod rot, Pythium spp., Rhizoctonia, seedling rot.

STREPTOMYCES GRISEOVIRIDES,　　　**Mfg: Kemira/Ag Bio**
MYCOSTOP

Fusarium, Alternaria, Botrytis, seedrots, rootrots, wilt, Phomopsis, gray mold, Pythium, Phytophtora.

STREPTOMYCIN, AGRIMYCIN,　　　　**Mfg: Merck Ag Vet**
AGRI-STREP

Bacterial blight, bacterial leaf rot, bacterial spot, bacterial stem rot, bacterial wilt, black leg, blue mold, crown gall, fire blight, halo blight, ice nucleating bacteria, soft rot, wildfire, YTD in citrus.

SULFUR　　　　　　　　　　　　　**Mfg: Numerous**

Almond mite, Atlantic mite, blister mite, brown rot, bud mite, cherry leaf spot, citrus leaf mite, citrus rust mite, European red mite, flatmite, flyspeck, leaf spot, Pacific mite, peach scab, Ptetrobia mite, powdery mildew, red spider mite, russet mite, rust, scab, Septoria leaf spot, silver mite, six-spotted mite, sooty blotch, spider mite, thrips, two-spotted mite, Vienna spider mite.

TCMTB, BUSAN-30, NUSAN　　　　**Mfg: Buckman Labs, Inc.**

Barley stripe, covered smut, Curvularia spp., damping-off, Fusarium spp., head blight, Helminthosporium spp., loose smut, Penicillium spp., Phythium, rhizoctonia, rust, scab bunt, seed decay, seedling blight, seedling rots, smut, soreshin, stinking smut.

TEBUCONAZOLE, FOLICUR　　　　　**Mfg: Miles**

White mold.

TERRAMYCIN, MYCOSHIELD　　　　**Mfg: Merck Ag Vet**

Bacterial spot, fire blight, lethal decline, lethal yellowing disease, pear decline, X-disease.

THIABENDAZOLE, ARBORTECT, MERTECT

Mfg: Merck Ag Vet

Anthracnose, basal rot, black rot, blue mold, blue mold rot, Botrytis, brown spot, bullseye rot, Cercosporella foot rot, cobweb, crown rot, Dutch elm disease, dollar spot, dry bubble, foot rot, frogeye leafspot, Fusarium spp., glume blotch, green mold, gray mold, narrow brown leaf spot, neck rot, Penicillium spp., pod blight, purple seed stain, rice blast, Sclerotinia rot, scurf, sheath blight, stem blight, stem end rot, stem rot, tuber rot, wet bubble, white mold.

THIOPHANATE-METHYL, FUNGO, TOPSIN-M, DOMAIN, CLEARY 3336

Mfg: W.A. Cleary and Elf-Atochem

Anthracnose, bitter rot, black knot, black rot, blossom blight, Botrytis fruit rot, Botrytis gray mold, Botrytis spp., brown leaf spot, brown patch, brown rot, Ceratocystis, Cercospera leaf spot, copper spot, coscomyces, Dendrophoma leaf blight, Diplocarpon leaf scorch, dollar spot, early and late blight, eye spot, flyspeck, foot rot, frogeye leaf spot, fruit brown rot, Fusarium blight, Fusarium patch, Gloessporium spp., gray mold, gummy stem blight, late blight, leaf blight, leaf scorch, leaf spot, Monilinia, peach scab, Penicillium spp., pod and stem blight, powdery mildew, purple seed stain, red thread, Rhizoctonia spp., scab, sooty blotch, strawbreaker, stripe smut, target spot, Thielaviopis spp., white rot, pink snow mold, summer patch, necrotic ring, diplodiatis blight.

THIRAM

Mfg: Numerous

Alternaria spp., apple blotch, banana fruit spots, basal rot, bitter rot, black pox, black rot, Botryosphaeria, Brooks spot, brown patch, brown rot, cedar apple rust, Cercospora early blight, crown rust, damping-off, dollar spot, early blight, flyspeck, fruit rot, gray mold, Helminthosporium leaf spot, late blight, leaf blight, leaf scorch, leaf spot, Rhizoctonia, Rhizopus rot, root rot, scab, scruf, seed decay, smut, surface mold of bananas, snow mold, sooty blotch, stem rot.

TOP COP (basic copper sulfate/ sulfur)

Mfg: Stoller/others

Leaf spot, rust, Heliminthosporium, septoria, downy mildew, powdery mildew, halo blight, alternaria leaf spot, Cercospora blight, bacterial blight, early and late blight, bollrot, Anthracnose, Phomopsis, purple blotch, blast, stem rot, leaf smut, sheath blight, pod and stem blight, brown spot, blue mold, bacterial speck and spot, black rot, shot hole, brown rot, fruit spot, Sigatoka disease, melanose, bunch rot, peacock spot, fireblight, leaf and cane rot, damping off, seedling diseases.

TRIADIMEFON, BAYLETON   Mfg: Miles Inc.

Anthracnose, barley scald, black rot, brown patch, copper spot, dollar spot, flower blight, Fusarium blight, Fusarium patch, gray snow mold, leaf blight, leaf blotch, leaf spot, petal blight, pine rust, powdery mildew, pink snow mold, red thread, rusts, Sclerotinia spp., snow mold, Southern blight, stripe smut, tip blight, summerpatch, leaf glume blotch, tan spot.

TRIADIMENOL, BAYTAN   Mfg: Miles Inc. and Gustafson

Barley stripe, bunt, covered kernel smut, flag smut, footrot, glume blotch, leaf rust, leaf spot, loose smut, powdery mildew, Septoria, smuts, stem, rust, speckled leaf blotch, stinking smut, stripe rust, take-all, Typhida rot, wet blotch.

TRICHODERMA HARZIANUM, BINAB-T   Mfg: E.R. Butts Inc.

Wood decay.

TRIFLUMIZOLE, TERRAGUARD, PROCURE   Mfg: Uniroyal

Cylindrocladium root and crown rot, root rots, leaf spot, petiole rot, powdery mildew, crown canker, scab, Helminthosporium, Rhizoctonia, rusts, scab.

TRIFORINE, FUNGINEX   Mfg: CIBA, Valent

Anthracnose, black spot, blossom blight, brown rot, leaf spot, powdery mildew, mummy berry, rust, scab, fruit rot.

VINCLOZOLIN, ORNALIN, RONILAN, CURALAN   Mfg: BASF

Botrytis blight, Botrytis rot, brown rot, Ciborinia, gray mold, Molinia, Ovulinia, Sclerotinia spp., Stromatinia, brown patch, dollar spot, Fusariumm patch, gray snow mold, leaf spots, melting out, red thread, pink patch.

VITAVAX-EXTRA (carboxin/   **Mfg: Gustafson/Uniroyal**
imazalil/TBZ)

Rhizoctoni, Heliminthosporium, Fusarium, seed blotch, net block, Septoria, loose smut, common bent.

ZIRAM   **Mfg: Elf-Atochem and others**

Anthracnose, apple scab, black rot, blossom blight, Botrytis spp., brown rot, bullseye rot, Cercospora early blight, evergreen blackberry canker, evergreen canker fruit brown rot, fruit rot, Fusiform rust, gray mold, leaf spot, mummy berry, peach leaf curl, pecan scab, petal blight, powdery mildew, rust, scab, Septoria late blight, shot hole, twig blight.

ZYBAN, (thiophanate-methyl/   **Mfg:Scotts**
mancozeb

Anthracnose, black spot, downy mildew, flower blight, leaf blight, powdery mildew, rust, scab, stem and twig blight.

NOTES

GLOSSARY

SPRAYER CALIBRATION

Calibration of your spraying equipment is very important. It should be done at least every other day of operation to insure application of the proper dosages. This is probably the most important step in your whole spraying operation since applying incorrect amounts may do much more damage than good.

If a lower rate is desired it may be obtained by increasing the speed, reducing the speed, increasing the pressure or changing to a larger nozzle or a combination of the three.

GENERAL CALIBRATION

I. Method I.
   A. Measure out 660 feet.
   B. Determine the amount of spray put out in traveling this distance at the desired speed.
   C. Use this formula:
   $$\text{gallons/acre} = \frac{\text{gallons used in 660 feet} \times 66}{\text{swath width in feet}}$$
   *D. Fill the tank with the desired concentration.

II. Method II.
   A. Fill spray tank and spray a specified number of feet.
   B. After spraying refill tank measuring the quantity of material needed for refilling.
   C. Use this formula:
   $$\text{gallons/acre} = \frac{43560 \times \text{gallons delivered}}{\text{swath length (ft.)} \times \text{swath width (ft.)}}$$
   *D. Fill the tank with the desired concentrate.

III. Method III.
   A. Measure 163 feet in the field.
   B. Time tractor in 163 feet. Make two passes to check accuracy.
   C. At edge of field adust pressure valve until you catch 2 pints (32 ounces) of spray in the same amount of time it took to run the 163 feet. Be sure tractor is at the same throttle setting. You are now applying 20 gallons per acre on a 20 inch nozzle boom spacing.
   D. For each inch of nozzle spacing on boom, increase time by 5% or reduce the volume by 5%.

---

*If you have calibrated your rig and it is putting out 37 gallons/acre, the required dosage is 4 pounds actual/acre. Therefore, for every 37 gallons of carrier (water, oil, etc.) in the spray tank you add 4 pounds of active material.

USEFUL FORMULAE

1. To determine the amount of active ingredient needed to mix in the spray tank.
   No. of gallons or pounds =
   $$\frac{\text{No. of acres to be sprayed x pounds active ingredient required per A}}{\text{pounds active ingredient per gallon or per pound}}$$

2. To determine the amount of pesticide needed to mix a spray containing a certain percentage of the active ingredient.
   No. of gallons or pounds =
   $$\frac{\text{gallons of spray desired x \% active ingredient wanted x 8.345}}{\text{pounds active ingredient per gallon or pound x 100}}$$

3. To determine the percent active ingredient in a spray mixture.
   Percent =
   $$\frac{\begin{array}{c}\text{pounds or gallons of concentrate used (not just active ingredient)}\\ \text{x \% active ingredient in the concentrate}\end{array}}{\text{gallons of spray x 8.345 (weight of water/gallon)}}$$

4. To determine the amount of pesticide needed to mix a dust with a given percent active ingredient.
   pounds material =
   $$\frac{\text{\% active ingredient wanted x pounds of mixed dust wanted}}{\text{\% active ingredient in pesticide used}}$$

5. To determine the size of pump needed to apply a given number of gallons/acre.
   pump capacity =
   $$\frac{\text{gallons/acre desired x boom width (feet) x mph}}{495}$$

6. To determine the nozzle capacity in gallons per minute at a given rate/acre and miles/hour.
   Nozzle capacity =
   $$\frac{\text{gallons/acre x nozzle spacing (inches) x mph}}{5940}$$

7. To determine the acres per hour sprayed.
   Acres per hour =
   $$\frac{\text{swath width (inches) x mph}}{100}$$

8. To determine the rate of speed in miles per hour.
   1. Measure off a distance of 300 to 500 feet.
   2. Measure in seconds the time it takes the tractor to go the marked off distance.
   3. Multiply .682 times the distance traveled in feet and divide product by the number of seconds.
   $$\text{MPH} = \frac{.682 \text{ x distance}}{\text{seconds}}$$

9. To determine the nozzle flow rate.
   Time the seconds necessary to fill a pint jar from a nozzle.
   Divide the number of seconds into 7.5.

   gallons/minute/nozzle = $\dfrac{7.5}{\text{seconds}}$

10. To determine the gallons per minute per boom.
    Figure out the gallons/minute/nozzle and multiply by the number of nozzles.

11. To determine the gallons per acre delivered.

    $$\frac{5940 \times \text{gallons/minute/nozzle}}{\text{nozzle spacing (inches)} \times \text{mph}} = \text{gpa}$$

12. To determine the acreage sprayed per hour.

    acres sprayed/hour = $\dfrac{\text{boom width (feet)} \times \text{mph}}{12}$

    This allows 30% of time for filling, turning, etc.

13. Sprayer Tank Capacity
    Calculate as follows:
    1. Cylindrical Tanks:
       Multiply the length in inches times the square of the diameter in inches and multiply the product by .0034.
       length x diameter squared x .0034 = number of gallons.
    2. Elliptical Tanks:
       Multiply the length in inches times the short diameter in inches times the long diameter in inches times .0034.
       length x short diameter x long diameter x .0034 = number of gallons.
    3. Rectangular Tanks:
       Multiply the length times the width times the depth in inches and multiply the product by .004329.
       length x width x depth x .004329 = number gallons.

14. To determine the acres in a given area.
    Multiply the length in feet times the width in feet times 23. Move the decimal point 6 places to the left to give the actual acres.

CONVERSION TABLES (U.S.)

Linear Measure —
    1 foot = 12 inches
    1 yard = 3 feet
    1 rod = 5.5 yards = 16.5 feet
    1 mile = 320 rods = 1760 yards = 5280 feet

Square Measure —
    1 square foot (sq. ft.) = 144 square inches (sq. in.)
    1 square yard (sq. yd.) = 9 sq. ft.
    1 square rod (sq. rd.) = 272.25 sq. ft. = 30.25 sq. yd.
    1 acre = 43560 sq. ft. = 4840 sq. yds. = 160 sq. rds.
    1 square mile = 640 acres

Cubic Measure —
    1 cubic foot (cu. ft.) = 1728 cubic inches (cu. in.) = 29.922 liquid
        quarts = 7.48 gallons
    1 cubic yard = 27 cubic feet

Liquid Capacity Measure —
    1 tablespoon = 3 teaspoons
    1 fluid ounce = 2 tablespoons
    1 cup = 8 fluid ounces
    1 pint = 2 cups = 16 fluid ounces
    1 quart = 2 pints = 32 fluid ounces
    1 gallon = 4 quarts = 8 pints = 128 fluid ounces

Weight Measure —
    1 pound (lb.) = 16 ounces (oz.)
    1 hundred weight (cwt.) = 100 pounds
    1 ton = 20 cwt. = 2000 pounds

Rates of Application —
    1 ounce/sq. ft. = 2722.5 lbs./acre
    1 ounce/sq. yd. = 302.5 lbs./acre
    1 ounce/100 sq. ft. = 27.2 lbs./acre
    1 pound/100 sq. ft. = 435.6 lbs./acre
    1 pound/1000 sq. ft. = 43.6 lbs./acre
    1 gallon/acre = 3 ounces/1000 sq. ft.
    5 gallons/acre = 1 pint/1000 sq. ft.
    100 gallons/acre = 2.5 gallons/1000 sq. ft. = 1 quart/100 sq. ft.
    100 lbs./acre = 2.5 lbs./1000 sq. ft.

Important Facts —
Volume of sphere = diameter ^{3} x .5236
Diameter = circumference x .31831
Area of circle = diameter ^{2} x .7854
Area of ellipse = product of both diameters x .7854
Volume of cone = area of base x 1/3 height
1 cubic foot water = 7.5 gallons = 62.5 pounds
Pressure in psi = height (ft.) x .434
1 acre = 209 feet square
ppm = % x 10,000
% = ppm ÷ 10,000
1% by volume = 10,000 ppm

TABLE OF CONVERSION FACTORS

To Convert From	To	Multiply By
Cubic feet	gallons	7.48
Cubic feet	liters	28.3
Gallons	milliliters	3785
Grams	pounds	.0022
Grams/liter	parts/million	1000
Grams/liter	pounds/gallon	.00834
Liters	cubic feet	.0353
Milligrams/liter	parts/million	1
Milliliters/gallons	gallons	.0026
Ounces	grams	28.35
Parts/million	grams/liter	.001
Parts/million	pounds/million gallons	8.34
Pounds	grams	453.59
Pounds/gallon	grams/liter	111.83

1 gram = .035 ounce
1 kilogram = 2.2 lbs.
1 quintal = 100 kg. = 221 lbs.
1 metric ton = 100 kg. = 2,205 lbs.
1 hectare = 2.5 acres
1 meter = 39.4 inches
1 kilometer = .6 mile

CONVERSION TABLE

1 kilogram (kg) = 1000 grams (g) = 2.2 pounds
1 gram (g) = 1000 milligrams (mg) = .035 ounce
1 liter = 1000 milliliters (ml) or cubic centimeters (cc) = 1.058 quarts
1 milliter or cubic centimeter = .034 fluid ounce
1 milliliter or cubic centimeter of water weighs 1 gram
1 liter of water weighs 1 kilogram
1 pound = 453.6 grams
1 ounce = 28.35 grams
1 pint of water weighs approximately 1 pound
1 gallon of water weighs approximately 8.34 pounds

1 part per million (ppm) = 1milligram/liter
                        = 1 milligram/kilogram
                        = .0001 percent
                        = .013 ounces in 100 gallons of water

1 percent = 10.000 ppm
          = 10 grams per liter
          = 10 grams per kilogram
          = 1.33 ounces by weight per gallon of water
          = 8.34 pounds/100 gallons of water

.1 percent = 1000 ppm = 1000 milligrams/liter
.01 percent = 100 ppm = 100 milligrams/liter
.001 percent = 10 ppm = 10 milligrams/liter
.0001 percent = 1 ppm = 1 milligram/liter

CHEMICAL ELEMENTS

Name	Symbol	Atomic Weight	Valance
Aluminum	Al	26.97	3
Antomony	Sb	121.76	3, 5
Arsenic	As	74.91	3, 5
Barium	Ba	137.36	2
Bismuth	Bi	209.00	3, 5
Boron	B	10.82	3, 0
Bromine	Br	79.916	1, 3, 5, 7
Cadmium	Cd	112.41	2
Calcium	Ca	40.08	2
Carbon	C	12.01	2, 4
Chlorine	Cl	35.457	1, 3, 5, 7
Cobalt	Co	58.94	2, 3
Copper	Cu	63.57	1, 2
Fluorine	F	19.00	1
Hydrogen	H	1.008	1
Iodine	I	126.92	1, 3, 5, 7
Iron	Fe	55.85	2, 3
Lead	Pb	207.21	2, 4
Magnesium	Mg	24.32	2
Mercury	Hg	200.61	1, 2
Molybdenum	Mo	95.95	3, 4, 6
Nickel	Ni	58.69	2, 3
Nitrogen	N	14.008	3, 5
Oxygen	O	16.00	2
Phosphorus	P	30.98	3, 5
Potassium	K	39.096	1
Selenium	Se	78.96	3
Silicon	Si	28.06	4
Silver	Ag	107.88	1
Sodium	Na	22.997	1
Sulfur	S	32.06	2, 4, 6
Thallium	Tl	204.29	1, 3
Tin	Sn	118.70	2, 4
Titanium	Ti	47.90	3, 4
Uranium	U	238.17	4, 6
Zinc	Zn	65.38	2

WIDTH OF AREA COVERED TO ACRES PER MILE TRAVELED

Width of Strip (feet)	Acres/mile
6	.72
10	1.21
12	1.45
12	1.45
16	1.93
18	2.18
20	2.42
25	3.02
30	3.63
50	6.04
75	9.06
100	12.1
150	18.14
200	24.2
300	36.3

TEMPERATURE CONVERSION TABLE RELATIONSHIP OF CENTIGRADE AND FAHRENHEIT SCALES

°C	°F	°C	°F
-40	-40	25	77
-35	-31	30	86
-30	-22	35	95
-25	-13	40	104
-20	-4	45	113
-15	5	50	122
-10	14	55	131
-5	23	60	140
0	32	80	176
5	41	100	212
10	50		
15	59		
20	68		

PROPORTIONATE AMOUNTS OF DRY MATERIALS

Water	Quantity of Material				
100 gallons	1 lb.	2 lbs.	3 lbs.	4 lbs.	5 lbs.
50 gallons	8 oz.	1 lb.	24 oz.	2 lbs.	2 1/2 lbs.
5 gallons	3 tbs.	1 1/2 oz.	2 1/2 oz.	3 1/4 oz.	4 oz.
1 gallon	2 tsp.	3 tsp.	1 1/2 tbs.	2 tbs.	3 tbs.

PROPORTIONATE AMOUNTS OF LIQUID MATERIALS

Water		Quantity of Material	
100 gallons	1 qt.	1 pt.	1 cup
50 gallons	1 pt.	1 cup	1/2 cup
5 gallons	3 tbs.	5 tsp.	2 1/2 tsp.
1 gallon	2 tsp.	1 tsp.	1/2 tsp.

MILES PER HOUR CONVERTED TO FEET PER MINUTE

MPH	fpm
1	88
2	176
3	264
4	352

EMULSIFIABLE CONCENTRATE PERCENT RATINGS IN POUNDS ACTUAL PER GALLON

%EC	lbs./Gallon
10-12	1
15-20	1.5
25	2
40-50	4
60-65	6
70-75	8
80-100	10

CONVERSION TABLE FOR LIQUID FORMULATIONS*

Concentration of Active Ingredient in Formulations, lbs./gal.

Rate Desired Lbs./A	1	2	2.5	3	4	5	6
			(ml of formulation per 100 square feet)				
1	8.67	4.33	3.47	2.89	2.17	1.73	1.44
2	17.3	8.67	6.93	5.78	4.33	3.47	2.89
3	26.0	13.0	10.4	8.67	6.50	5.20	4.33
4	34.8	17.4	13.9	11.6	8.69	6.95	5.79
5	43.4	21.7	17.4	14.5	10.0	8.68	7.24
6	52.1	26.0	20.8	17.4	13.0	10.4	8.68
7	60.8	30.4	24.3	20.3	15.2	12.2	10.1
8	69.4	34.7	27.8	23.1	17.4	13.9	11.6
9	78.1	39.0	31.2	26.0	19.5	15.6	13.0
10	86.7	43.3	34.7	28.9	21.7	17.3	14.4

*Example: To put out a 100 sq. ft. plot at the rate of 5 lbs./A active ingredient using a formulation containing 4 lbs./gal. active ingredients, use 10.7 ml. of the 4 lbs./gal. formula and distribute evenly.

CONVERSION TABLE FOR DRY FORMULATIONS

Concentration of Active Ingredient in Formulation

Rate Desired Lbs./A	100%	90%	80%	75%	70%	60%	50%	40%	30%	25%	20%	10%	5%
	(Grams of formulation per 100 square feet)												
1	1.04	1.16	1.30	1.39	1.49	1.74	2.08	2.60	3.47	4.17	5.21	10.4	20.8
2	2.08	2.31	2.60	2.78	2.98	3.47	4.17	5.21	6.94	8.33	10.4	20.8	41.7
3	3.12	3.47	3.90	4.17	4.46	5.20	6.25	7.81	10.4	12.5	15.6	31.2	62.5
4	4.17	4.63	5.21	5.55	5.95	6.94	8.33*	10.4	13.9	16.7	20.8	41.7	83.3
5	5.21	5.79	6.51	6.94	7.44	8.68	10.4	13.0	17.4	20.8	26.0	52.1	104
6	6.25	6.94	7.81	8.33	8.93	10.4	12.5	15.6	20.8	25.0	31.2	62.5	125
7	7.29	8.10	9.11	9.72	10.4	12.1	14.6	18.2	24.3	29.2	36.4	72.9	146
8	8.33	9.26	10.4	11.1	11.9	13.9	16.7	20.8	27.8	33.3	41.7	83.3	167
9	9.37	10.4	11.7	12.5	13.4	15.6	18.7	23.4	31.2	37.5	46.9	93.7	187
10	10.4	11.6	13.0	13.9	14.9	17.4	20.8	26.0	34.7	41.7	52.1	104	208

*Example: To put out a 100 sq. ft. plot at the rate of 4 lbs./A active ingredient using a formulation containing 50% active ingredient, use 8.33 grams of the 50% formulation and distribute evenly over the 100 sq. ft.

CONVERSION TABLE FOR GRANULR FORMULATIONS

Concentration of Active Ingredient in Formulation

Rate Desired Lbs./A	20%	15%	10%	7.5%	5%	4%	3%	2%	1%
1	5.2	6.94	10.4	13.86	20.8	26.0	34.66	52.0	104.0
2	10.4	13.9	20.8	27.7	41.7	52.0	69.3	104.0	208.0
3	15.6	20.8	31.2	41.6	62.5	78.0	103.9	156.0	312.0
4	20.8	27.8	41.7	55.4	83.3	104.0	138.6	208.0	416.0
5	26.0	34.7	52.1	69.3	104.0*	130.0	173.3	260.0	520.0
6	31.2	41.6	62.5	83.2	125.0	156.0	207.9	312.0	624.0
7	36.4	45.6	72.9	97.0	146.0	182.0	242.6	364.0	728.0
8	41.7	55.5	83.3	110.9	167.0	208.0	277.3	416.0	832.0
9	46.9	62.5	93.7	124.7	187.0	234.0	311.9	468.0	936.0
10	52.1	69.4	104.0	168.6	208.0	260.0	346.6	520.0	1040.0
15	78.0	104.1	156.0	207.9	312.0	390.0	519.2	780.0	1560.0
20	104.0	138.8	208.0	277.2	416.0	520.0	693.2	1040.0	2080.0
25	130.0	173.5	260.0	346.5	520.0	650.0	866.5	1300.0	2600.0
30	156.0	208.2	312.0	415.8	624.0	780.0	1039.8	1560.0	3120.0

*Example: To put out a 100 sq. ft. plot at the rate of 5 lbs./A active ingredient using a formulation containing 5% active material, use 104.0 grams of the 5% formulation and distribute it evenly over the 100 sq. ft.

GRAMS/GALLON TABLE

Gallons PPM	5	10	15	20	25	50	75	100	150	200	300	400
5	0.1	0.2	0.3	0.4	0.5	1.0	1.4	1.9	2.8	3.8	5.7	7.6
10	0.2	0.4	0.6	0.8	1.0	1.9	2.8	3.8	5.7	7.6	11.0	15.0
15	0.3	0.6	0.9	1.1	1.4	2.8	4.3	5.7	8.5	11.0	17.0	23.0
20	0.4	0.8	1.1	1.5	1.9	3.8	5.7	7.6	11.0	15.0	23.0	30.0
25	0.5	0.9	1.4	1.9	2.4	4.7	7.1	9.5	14.0	19.0	28.0	38.0
50	0.9	1.9	2.8	3.8	4.7	9.5	14.0	19.0	28.0	38.0	57.0	76.0
75	1.4	2.8	4.3	5.7	7.1	14.0	21.0	28.0	43.0	57.0	85.0	114.0
100	1.9	3.8	5.7	7.6	9.5	19.0	28.0	38.0	57.0	76.0	114.0	151.0
125	2.4	4.7	7.1	9.5	12.0	24.0	36.0	47.0	71.0	95.0	142.0	189.0
150	2.8	5.7	8.5	11.0	14.0	28.0	43.0	57.0	85.0	114.0	170.0	227.0
175	3.3	6.6	9.9	13.0	17.0	33.0	50.0	66.0	99.0	133.0	199.0	265.0
200	3.8	7.6	11.0	15.0	19.0	38.0	57.0	76.0	114.0	151.0	227.0	303.0
250	4.7	9.5	14.0	19.0	24.0	47.0	71.0	95.0	142.0	189.0	284.0	379.0
300	5.7	11.0	17.0	23.0	28.0	57.0	85.0	114.0	170.0	227.0	341.0	454.0
400	7.6	15.0	23.0	30.0	38.0	76.0	114.0	151.0	227.0	303.0	454.0	606.0

DETERMINE THE NUMBER OF ROWS TO THE ACRE

Length of Rows

Rows/Acre	32"	36"	38"	40"	60"
1	16335	14520	13756	13068	8712
2	8168	7260	6878	6534	4356
3	5445	4840	4585	4356	2904
4	4084	3630	3439	3267	2178
5	3267	2904	2751	2614	1742
6	2723	2420	2293	2178	1452
7	2334	2074	1965	1867	1245
8	2042	1815	1719	1634	1089
9	1815	1613	1528	1452	968
10	1634	1452	1376	1307	871
11	1485	1320	1251	1188	792
12	1361	1210	1156	1089	726
13	1257	1117	1058*	1005	670
14	1167	1037	982	933	622
15	1089	968	917	871	581
16	1021	908	760	817	545
17	961	854	809	769	512
18	908	807	764	726	484
19	860	764	724	688	459
20	817	726	688	653	436
21	778	691	655	622	415
22	743	660	625	594	396
23	710	631	598	568	379
24	681	605	573	544	363
25	653	581	550	523	348
26	628	558	529	503	335
27	605	538	509	484	323
28	583	519	491	467	311
29	563	501	474	450	300
30	545	484	459	436	290

*Example: A grower's field is 1058 feet long furrowed out on 38-inch centers. Therefore, every 13 rows across the field represents an acre.

174

QUICK CONVERSIONS

TEMPERATURE			LENGTH			VOLUME	
°C	°F		cm	inch		liters	quarts
100	212		2.5	1		1	1.1
90	194		5	2		2	2.1
80	176		10	4		3	3.2
70	158		20	8		4	4.2
60	140		30	12		5	5.3
50	122		40	16		6	6.3
40	104		50	20		7	7.4
35	95		60	24		8	8.5
30	86		70	28		9	9.5
25	77		80	32			
20	68		90	36			
15	59		100	39			
10	50		200	79			
5	41			feet			
0	32		300	10			
-5	23		400	13			
-10	14		500	16			
-15	5		1,000	33			
-20	-4						
-25	-13						
-30	-22						
-40	-40						

QUICK CONVERSIONS

kg./ha.	lb./A
1	0.9
2	1.8
3	2.7
4	3.6
5	4.5
10	9
20	18
20	27
40	36
50	45
60	54
70	62
80	71
90	80
100	89
200	180
300	270
400	360
500	450
600	540
700	620
800	710
900	800
1000	890
2000	1800

kg./ha.	ton/A
3000	1¼
4000	1¾
5000	2¼
6000	2¾
7000	3
8000	3½
9000	4
10000	4½
11000	5
12000	5½
13000	5¾
14000	6¼
15000	6¾
16000	7
17000	7½
18000	8
19000	8½
20000	9

USEFUL MEASUREMENTS

LENGTH

1 mile = 80 chains = 8 furlongs = 1760 yards = 5280 feet = 1.6 kilometers
1 chain = 22 yards = 4 rods, poles or perches = 100 links

WEIGHT

1 long ton = 20 cwt. = 2240 pounds
1 pound = 16 ounces = 454 grams = 0.454 kilograms
1 short ton = 2000 pounds
1 metric ton = 2204 pounds = 1000 kilograms

AREA

1 acre = 10 sq. chains = 4840 sq. yards = 43560 sq. ft. = 0.405 hectares
1 sq. mile = 640 acres = 2.59 kilometers
1 hectare = 2.471 acres

VOLUME

1 gal. = 4 quarts = 8 pints = 128 fluid ozs. = 3.785 liters
1 fluid oz. = 2 tablespoons = 4 dessertspoons = 8 teaspoons = 28.4 c.c.'s

CAPACITIES

Cylinder — diameter $^2/$ x depth x 0.785 = cubic feet
Rectangle — breadth x depth x length = cubic feet
Cubic feet x 6.25 = gallons

QUICK CONVERSIONS

1 pint/acre	= 1 fluid oz./242 sq. yards
1 gal./acre	= 1 pint/605 sq. yards
1 lb./acre	= 1 oz./300 sq. yards
1 cwt./acre	= 0.37 oz./sq. yard
1 m.p.h.	= 88 ft./minute
3 m.p.h.	= 1 chain/15 sec.
1 liter/hectare	= 0.089 gal./acre
1 kilogram/hectare	= 0.892 lb./acre
1 c.c./100 liters	= 0.16 fl. oz./100 gallons
125 c.c./100 liters	= 1 pint/100 gallons
1 gm./100 liters	= 0.16 oz./100 gallons

A strip 3 ft. wide x 220 chains
A strip 4 ft. wide x 165 chains } 1 acre
A strip 5 ft. wide x 132 chains

CONVERSION FACTORS USED IN CALCULATION

Convert	To	By
gram (gm.)	kilogram (kg.)	move decimal 3 places to left

Example: 2000 gm. = 2.0 kg.

gram (gm.)	milligram (mg.)	move decimal 3 places to right

Example: 2.0 gm. = 2000 mg.

gram (gm.)	pound (lb.)	divide by 454

Example: 658 gm./ ÷ 454 = 1.45 lb.

gram/pound	percent (%)	divide by 4.54

Example: 90 gm./lb. ÷ 4.54 = 19.8%

gram/ton	percent	multiply by 11, move decimal

Example: 45 gm./ton x 11 = 495 = .00495%　　　　5 places to left

kilogram (kg.)	gram	move decimal 3 places to right

Example: 5.5 kg. = 5500 gm.

milligram (mg.)	gram	move decimal 3 places to left

Example: 95 mg. = 0.095 gm.

percent	gram/pound	multiply by 4.54

Example: 25 x 4.54 = 113.5 gm./lb.

percent	parts/million (ppm)	move decimal 4 places to right

Example: .025% = 250 ppm.

percent	gram/ton	divide by 11, move

Example: .011 ÷ 11 = .001 = 100 gm./ton　　　　decimal 5 places to right

pound	gram	multiply by 454

Example: 0.5 lb. x 454 = 227 gm.

ppm	percent	move decimal 4 places to left

Example: 100 ppm = 0.01%

STANDARD MEASUREMENTS

MEASURE OF LENGTH (Linear Measure)

4 inches	=	1 hand
9 inches	=	1 span
12 inches	=	1 foot
3 feet	=	1 yard
6 feet	=	1 fathom
5½yards - 16½ feet	=	1 rod
40 poles	=	1 furlong
8 furlongs	=	1 mile
5,280 feet = 1,760 yards	=	320 rods = 1 mile
3 miles	=	1 league

MEASURE OF SURFACE (area)

144 square inches	=	1 square foot
9 square feet	=	1 square yard
30¼ square yards	=	1 square rod
160 square rods	=	1 acre
43,560 square feet	=	1 acre
640 square acres	=	1 square mile
36 square miles	=	1 township

SURVEYOR'S MEASURE

7.92 inches	=	1 link
25 links	=	1 rod
4 rods	=	1 chain
10 square chains	=	160 square rods = 1 acre
640 acres	=	1 square mile
80 chains	=	1 mile
1 Gunter's chain	=	66 feet

METRIC LENGTH

1 inch	=	2.54 centimeters
1 foot	=	.305 meter
1 yard	=	.914 meter
1 mile	=	1.609 kilometers
1 fathom	=	6 feet
1 knot	=	6,086 feet
3 knots	=	1 league
1 centimeter	=	.394 inch
1 meter	=	3.281 feet
1 meter	=	1.094 yards
1 kilometer	=	.621 mile

TROY WEIGHT

24 grains	=	1 pennyweight
20 pennyweight	=	1 ounce
12 ounces	=	1 pound

LIQUID MEASURE

2 cups	=	1 pint
4 gills	=	1 pint
16 fluid ounces	=	1 pint
2 pints	=	1 quart
4 quarts	=	1 gallon
31½ gallons	=	1 barrel
2 barrels	=	1 hogshead
1 gallon	=	231 cubic inches
1 cubic foot	=	7.48 gallons
1 teaspoon	=	.17 fluid ounces (1/6 oz.)
3 teaspoons (level)	=	1 tablespoon (1/2 oz.)
2 tablespoons	=	1 fluid ounce
1 cup (liquid)	=	16 tablespoons (8 oz.)
1 teaspoon	=	5 to 6 cubic centimeters
1 tablespoon	=	15 to 16 cubic centimeters
1 fluid ounce	=	29.57 cubic centimeters

CUBIC MEASURE (Volume)

1,728 cubic inches	=	1 cubic foot
27 cubic feet	=	1 cubic yard
2,150.42 cubic inches	=	1 standard bushel
231 cubic inches	=	1 standard gallon (liquid)
1 cubic foot	=	4/5 of a bushel
128 cubic feet	=	1 cord (wood)
7.48 gallons	=	1 cubic foot
1 bushel	=	1.25 cubic feet

DRY MEASURE

2 pints	=	1 quart
8 quarts	=	1 peck
4 pecks	=	1 bushel
36 bushels	=	1 chaldron

APOTHECARIES' WEIGHT

20 grains	=	1 scruple
3 scruples	=	1 dram
8 drams	=	1 ounce
12 ounces	=	1 pound
27-11/32 grains	=	1 dram
16 drams	=	1 ounce
16 ounces	=	1 pound
2,000 pounds	=	1 ton (short)
2,240 pounds	=	1 ton (long)

CONVERSION FACTORS

Degree C = 5/9 (Degree F − 32).

Degree F = 9/5 (Degree C + 32).

Degrees Absolute (Kelvin) = Degrees centigrade + 273.16.

Degrees absolute (Rankine) = Degrees fahrenheit + 459.69.

Multiply	By	To Obtain
Diameter circle	3.1416	Circumference circle
Diameter circle	0.8862	Side of equal square
Diameter circle squared	0.7854	Area of circle
Diameter sphere squared	3.1416	Area of sphere
Diameter sphere cubed	0.5236	Volume of sphere
U.S. Gallons	0.8327	Imperial gallons (British)
U.S. Gallons	0.1337	Cubic feet
U.S. Gallons	8.330	Pounds of water (20° C)
Cubic feet	62.427	Pounds of water (4° C)
Feet of water (4° C)	0.4335	Pounds per square inch
Inch of mercury (0° C)	0.4912	Pounds per square inch
Knots	1.1516	Miles per hour

Figuring Grain Storage Capacity

1 bu. ear corn = 70 lbs. 2.5 cu. ft. (15.5% moisture)

1 bu. shelled corn = 56 lbs. 1.25 cu. ft. (15.5% moisture)

1 cu. ft. = 1/2.50 = .4 bu. of ear corn

1 cu. ft. = 1/1.25 = .8 bu. of shelled corn; Bu. x 1.25 ft.³, ft.³ x .8 bu.

Ft.³ = Bu. x 1.25

Bu. = Ft.³ x .8

Rectangular or square cribs or bins

    cu. ft. = width x height x length (W x H x L)

Round cribs, bins or silos (= 3.1416)

    Volume = R²H = D²H/r

    cu. ft. = x radius x radius x height = (R x R x H)

    or x diameter x diameter x height = (D x D x H)

$$\frac{\text{or } 3.1416 \times D \times D \times H}{4} = .785 \times D \times D \times H$$

Examples

1. Crib — ear corn — 6' wide by 12' high by 40' long
   a. 6 x 12 x 40 = 2880 cu. ft. x 4 bu./cu. ft. = 1152 bu.
   b. 6 x 12 x 1 = 72 cu. ft. x .4 28.8 bu./ft. of length x 40' = 1152 bu.

2. Round crib — ear corn — 14' diameter by 14' high
   a. .785 x 14' x 14' x 14' x .8 = 1722 bushel
   b. .785 x 14' x 1 x .4 6.15 bu./ft. x 14 = 861 bushel

3. Round Bin or Silo — shell corn — 14' diameter by 14' high
   a. .785 x 14' x 14' x 14' x .8 = 1722 bushel
   b. .785 x 14' x 14' 14' x 1 x .8 - 123 bu./ft. x 14' = 1722 bushel

Metric Weight

1 grain	=	.065 gram
1 apothecaries' scruple	=	1.296 grams
1 avoirdupois ounce	=	28.350 grams
1 troy ounce	=	31.103 grams
1 avoirdupois pound	=	.454 kilogram
1 troy pound	=	.373 kilogram
1 gram	=	15.432 grains
1 gram	=	.772 apothecaries' scruple
1 gram	=	.035 avoirdupois ounce
1 gram	=	.032 troy ounce
1 kilogram	=	2.205 avoirdupois pounds
1 kilogram	=	2.679 troy pounds

Capacity

1 U.S. fluid ounce	=	29.573 ml
1 U.S. fluid quart	=	.946 liter
1 U.S. fluid ounce	=	29,573 milliliters
1 U.S. liquid quart	=	.964 liter
1 U.S. dry quart	=	1.101 liters
1 U.S. gallon	=	3.785 liters
1 U.S. bushel	=	.3524 hectoliters
1 cubic inch	=	16.4 cubic centimeters
1 liter	=	1,000 milliliters or 1,000 cubic centimeters
1 cubic foot water	=	7.48 gallons or 62-1/2 pounds
231 cubic inches	=	1 gallon
1 milliliter	=	.034 U.S. fluid ounce
1 liter	=	1.057 U.S. liquid quarts
1 liter	=	.908 U.S. dry quart
1 liter	=	.264 U.S. gallon
1 hectoliter	=	2.838 U.S. bushels
1 cubic centimeter	=	.061 cubic inch

Miscellaneous Equivalents

9 in. equals 1 span

6 ft. equals 1 fathom

6,080 ft. equals 1 nautical mile

1 board ft. equals 144 cu. in.

1 cylindrical ft. contains 4-7/8 gals.

1 cu. ft. equals .8 bushel

12 dozen (doz.) equals 1 gross (gr.)

1 gal. water weighs about 8-1/3 lbs.

1 gal. milk weighs about 8.6 lbs.

1 gal. cream weighs about 8.4 lbs.

46-1/2 qts. of milk weighs 100 lbs.

1 cu. ft. water weighs 62-1/2 lbs., contains 7-1/2 gals.

1 gal. kerosene weighs about 6-1/2 lbs.

1 bbl. cement contains 3.8 cu. ft.

1 bbl. oil contains 42 gals.

1 standard bale cotton weighs 480 lbs.

1 keg of nails weighs 100 lbs.

4 in. equals 1 hand in measuring horses

ADDRESSES OF
BASIC MANUFACTURERS

Abbott Laboratories
Abbott Park-1400 Sheridan Road
N. Chicago, IL 60064

Agbiochem Inc.
3 Fleetwood Court
Orinda, CA 94563

AgriDyne Technologies
2401 Foothill Dr.
Salt Lake City, Utah 84109-1405

Agrigenetics Co.
5649 E. Buckeye Road
Madison, WI 53716

AgrEvo
Postfach 270654
13476 Berlin, Germany

AgrEvo
Postfach 80 03 20
6230 Frankfurt (m) 80
Germany

AgrEvo
3509 Silverside Rd.
Wilmington, DE 19803

Agro-Kanesho Co. Ltd.
1-1-Marunonchi-1-Chome
Tokyo 100 Japan

Agrolinz
Agrarchemihalein GmbH
St. Peter Strasse 25
POB 21
A-4021 Linz, Austria

Agtrol Chemical Products
7322 Southwest Freeway,
Ste 1400
Houston, TX 77074

Agway Inc.
P.O. Box 4933
Syracuse, NY 13221-4933

Albaugh Chemical Corp.
1571 Ankery Blvd. Ste. A
Ankeny, IA 50021

American Cyanamid Co.
One Cyanamid Plaza
Wayne, NJ 07470

Amvac Chemical Corp.
4100 E. Washington Blvd.
Los Angeles, CA 90023

The Andersons
P.O. Box 119
Maumee, OH 43537

Applied Biochemists
6120 W. Douglas Ave.
Milwaukee, WI 53218

Asahi Chemical Industry Co.
500 Oaza-Takayasu
Ikarugacho, Ikomagun
Nara 636-01 Japan

Avitrol Corp.
7644 E. 46th St.
Tulsa, OK 74145

Bactec Corp.
9601 Katy Freeway Ste. 350
Houston, TX 77024-1333

BASF Akfiengesellschaft
Crop Protection
Postfach 120
D-67114 Limburgerhof,
Germany

BASF
Ag Chemical Div.
P.O. Box 13528
Research Triangle Park, NC 27709

Bayer AG
Crop Protection
D-51368 Leverkusen
Bayerwerk, Germany

Bell Labs
3699 Kinsman Boulevard
Madison, WI 53704

Biocontrol Ltd.
719 Second St., Suite 12
Davis, CA 95616

Biologic Inc.
11 Lake Ave. Extension
Danbury, Conn. 06811

Biosys
1057 E. Meadow Circle
Palo Alto, CA 94303

Bio Technica Inlt. Corp.
7300 W. 110th St. #540
Overland Park, KS 66210

Boehringer Ingelheim Animal Health
2621 N. Belt Highway
St. Joseph, MO 64506

Boliden Intertrade Inc.
3379 Peachtree Rd. NE
Suite 300
Atlanta, GA 30326

Brandt Consolidated
P.O. Box 277
Pleasant Plains, IL 62677

Brogdex Co.
1441 W. 2nd St.
Pomona, CA 91766

Buckman Labs
1256 N. McLean Blvd.
Memphis, TN 38108

Burlington Bio Medical Corp.
222 Sherwood Ave.
Farmingdale, NY 11735-1718

C.P. Chemicals, Inc.
1 Parker Place
Ft. Lee, NJ 07224

CFPI Agro Chem Inc.
20 Blvd. Cameliant
92233 Gennevillers, France

Calliope S.A.
BP 80
64150 Nogueres, France

Cedar Chemical Co.
5100 Poplar Ave.
24th Floor
Memphis, TN 38137

Cenex/Land O' Lakes
P.O. Box 64089
St. Paul, MN 55164-0089

Cheminova Agro
P.O. Box 9
DK-7620 Lemvig, Denmark

Cheminova Inc.
Oak Park Hill
1700 Rte 23 Ste 210
Wayne, NJ 07470

Chemol Trading Ltd. Co.
H-1134 Budapest XIII
Robert Karoly Korut 61-65,
Hungary

Chinoin - Agro Chemical
H-1780 Budapest
P.O. Box 49,
Hungary

CIBA-Geigy AG
CH-4002 Basel 7,
Switzerland

CIBA-Geigy Corporation
Agricultural Chemicals
P.O. Box 18300
Greensboro, NC 27419-8300

Cochran Corp.
P.O. Box 14603
Memphis, TX 38114-0603

W. A. Cleary Company
Southview Industrial Park
178 Rte 522 Ste A
Dayton, NJ 08810

Concep Inc.
213 SW Columbia
Bend, OR 97703

Coopers Animal Health
2000 S. 11th St.
Kansas City, KS 66103-1438

Cornbelt Chemical Co.
P.O. Box 410
McCook, NE 69001

Crop Genetics Intl.
10150 Old Colombia Road
Colombia, MD 21046-1704

CuproQuim Corp.
P.O. Box 171357
Memphis, TN 38187-1357

Cyanamid Forschung
Rechnugspru fung
Postfach 100
D-55270 Schwabenheim,
Germany

DK International
1000 Johnson Ferry Road C-110
Marietta, GA 30068

DNA Plant Technology Corp.
6701 San Pablo Ave.
Oakland, CA 94608

Dainippon Ink & Chemicals
DIC Building
7-20 Nihonbashi 3 Chome
Chuo-ku
Tokyo 103 Japan

Degesch America, Ind.
P.O. Box 116
Weyers Cave, VA 24486

Denka Intl. Inc.
P.O. Box 337
3771 NG Barneveld,
The Netherlands

Detia Degesch
Dr. Werner Freyberg Strasse 1
6947 Laudenbach Bergstrasse,
Germany

Dow Elanco
9330 Zionsville Rd.
Indianapolis, IN 46268

Dr. R. Maag Ltd.
Chemical Works
CH-8157
Dielsdorf, Switzerland

Drexel Chemical Co.
P.O. Box 9306
Memphis, TN 38109-0306

Dunhill Chemical Co.
3026 Muscatel Ave.
Rosemead, CA 91770

DuPont Company
Agricultural Chemical Dept.
Barley Mill Plaza
P.O. Box 80038
Wilmington, DE 19880-0038

J. T. Eaton & Co. Inc.
1393 E. Highland Road
Twinsburg, OH 44087

Ecogen
2005 Cabot Blvd., West
Langhorn, PA 19047-1810

Eco Science
1 Innovation Dr.
Worcester, MA 01605

Elf Atochem Agri
P.O. Box 6030
3196 XH Vondelingen Plaat,
The Netherlands

Elf Atochem North America
Ag Chemical Division
2000 Market St.
Philadelphia, PA 19103-3222

Endura S.P.A.
Viale Pietramellara 5
40121 Bologna, Italy

Enichem Agricoltura
Via Medici Del Vascella 40C
21038 Milano, Italy

FMC Corporation
Ag Chemical Div.
1735 Market Street
Philadelphia, PA 19103

Fair Products Inc.
P.O. Box 386
Cary, NC 27512-0386

Farmland Industries
P.O. Box 7305
Kansas City, MO 64116-0005

Fermenta Animal Health Co.
10150 N. Executive Hills Blvd.
Kansan City, MO 64153-2315

Gowan Co.
P.O. Box 5569
Yuma, AZ 85366-5569

W.R. Grace & Co.
7379 Rte. 32
Columbia, MD 21044

Great Lakes Chemical Corp.
P.O. Box 2200
W. Lafayette, IN 47906

Griffin Corporation
P.O. Box 1847
Valdosta, GA 31601

Gustafson, Inc.
1400 Preston Rd., Ste. 400
Plano, TX 75093

Hacco Inc.
P.O. Box 7190
Madison, WI 53707

Helena Chemical Co.
6075 Poplar Ave.
Suite 500
Memphis, TN 38119

Hendrix & Dail Inc.
P.O. Box 648
Greenville, NC 27835-0648

Hercon Environmental Co.
Aberdeen Road
Emigsville, PA 17318

Hess & Clark
7th and Orange Sts.
Ashland, OH 44805

Hodogaya Chemical
1-4-2 Toranomon-1-Chome
Minato-ku
Tokyo 105 Japan

Hokko Chemical Industries
Mitsui Building 2
4-4-20 Nihonbashi
Hongoku-cho 4-chone
Tokyo 103 Japan

Ihara Chemical Co.
1-4-26 Ikenohata 1-Chome
Taitoku
Tokyo 110 Japan

Ishihara Sangyo AgroLtd.
10-30 Fujimi 2-Chome
Chiyoda-ku, Tokyo 102
Japan

ISK Biosciences
5966 Heisley Rd.
Mentor, OH 44061-8000

Janssen Pharmaceutical
Plant Protection Div.
1125 Trenton-Harbouton Rd.
Titusville, NJ 08560

Janssen Pharmaceutica N. V.
Agricultural Div.
B-2340 Beerse, Belgium

Kaken Pharmaceutical Co., Ltd.
Mitsuihoncho Bldg.
4-10 Nihonbashi honcho 3-chone
Chuo-ku
Tokyo 103 Japan

Kemira Biotech
P.O. Box 330
00101 Helsinki, Finland

Kemira Oy
P.O. Box 44
SF-02271 Espoo, Finland

Kincaid Enterprises
P.O. Box 549
Nitro, WV 25143

Kumiai Chemical Industries
4-26 Ikenohata 1-Chome
Tokyo 110 Japan

Kureha Chemical Ind. Co.
1-9-11 Nihonbashi,
Horidome-cho, Chuo-ku,
Tokyo 103 Japan

Lebanon Agri Corp.
P.O. Box 180
Lebanon, PA 17042-0180

LESCO Inc
20005 Lake Road
Rocky River, OH 44116

Lipha Tech
34 Rue Saint Romain
69008 Lyon, France

Lipha Tech
3101 W. Custer Ave.
Milwaukee, WI 53209

Loveland Industries
P.O. Box 1289
Greeley, CO 80632

Luxenbourg Industries
P.O. Box 13
Tel Aviv 61000 Israel

Luxembourg-Pamol Inc.
5100 Poplar Ave. Ste. 2746
Memphis, Tenn. 38137

Maag Agrochemicals
P.O. Box 6430
Vero Beach, FL 32961-6430

MAGNA-Herbicide Div.
P.O. Box 11192
Bakersfield, CA 93389

A. H. Marks & Co.
Wyke Bradford
West Yord BD12 9EJ England

Maktheshim-Agan
551 Fifth Ave., Ste. 1100
New York, NY 10176

Maktheshim-Agan
P.O. Box 60
84100 Beer-Sheva, Israel

McLaughlin Gromley King
8810-10th Ave., North
Minneapolis, MN 55427

Meiji Seika Company
4-16 Kyobashi 2-Chome
Chuo-ku
Tokyo 104 Japan

Merck-Ag Vet
P.O. Box 2000
Rahway, NJ 07065-6430

E. Merck A. G.
61 D Armstadt
Franfurter Strasse 250, Germany

Micro-Flo Co.
P.O. Box 5948
Lakeland, FL 33807

Miles Inc.
P.O. Box 4913
Kansas City, MO 64120-6013

Miles Animal Health
P.O. Box 390
Shawnee Mission, KS 66201-0390

Miller Chemical & Fert. Corp.
Box 333
Hanover, PA 17331

Minerals Res. & Devel. Corp.
1Woodlawn Green
Suite 232
Charlotte, NC 28217

Mitsubishi Kasei Corp.
Agric. Chemical Div.
Mitsubishi Shozi Bldg.
5-2, Marunouchi 2-Chome
Chiyoda-ku
Tokyo 100 Japan

Mitsui Agricultural Chemicals
1-2-1 Ohtemachi, Chiyoda-ku,
Tokyo 100 Japan

Mitsui Toatsu Chemicals
Kasumigaschi Building
2-5 Kasamigaseki 3-chome
Chuyoda-hu
Tokyo 100 Japan

Monsanto Agricultural Group
800 N. Lindburgh Blvd.
St. Louis, MO 63167

Monterey Chemical Co.
P.O. Box 5317
Fresno, CA 93755

Motomco Ltd.
29 N. Ft. Harrison Ave.
Clearwater, FL 33615

MTM Agro Chemicals Ltd.
18 Liverpool Road, Great Sankey
Warrington, Cheshire, England

Mycogen Corp.
4980 Quail Canyon Rd.
San Diego, CA 92121

Nihon Bayer Agrochem
Itopia Nihonbashi Honcho Bldg.
7-1 Nihonbashi Honcho 2-chome
Chuo-hu
Tokyo 103 Japan

Nihon Nohyaku Company, Ltd.
2-5 Nihonbashi 1-Chome
Chuo-ku
Tokyo 103 Japan

Nippon Kayaku Co.-Ag Div.
Tokyo Kaijo Bldg.
2-1, Marunouchi 1-Chome
Chiyoda-ku
Tokyo 100 Japan

Nippon Soda Co., Ltd.
New-Ohtemachi Bldg.
2-1, 2-Chome Ohtemachi
Chiyoda-ku
Tokyo 100 Japan

Nissan Chemical Ind., Ltd.
Kowa-Hitotsubashi Bldg.
7-1, 3-Chome, Kanda
Nishiki-cho Chiyoda-ku
Tokyo 101 Japan

Nomix Inc.
300 Technology Ct. Ste. 50
Smyma, GA 30082

Nordox A/S
Ostensjoveien 13
0661 Oslo 6 Norway

Novo Bio Kontrol
Plant Protection Div.
33 Turner Road
Danbury, CT 06813-1907

Novo Bio Kontrol
Plant Protection Div.
Novo Alle
DK-2880 Bagsuaerd
Denmark

NPP S.A.
Routed Antix
BP 80
64150 Nogueres, France

NuFarm Ltd.
103-105 Pipe Rool
Laverton North
Victoia 3026, Australia

Old Bridge Chemicals
P.O. Box 194
Old Bridge, NJ 08857

Olympic Chemical Co.
P.O. Box K
Mainland, PA 19451

Ore-Calif. Chemicals Inc.
29454 Meadow View Rd.
Junction City, OR 97448

Otsuka Chemical Co.
2-27 Ohtedohir 3-chome
Chuo-ku
Osaka 540 Japan

PBI/Gordon Corp.
P.O. Box 4090
Kansas City, MO 64101

PMC Specialties Group
501 Murry Road
Cincinnati, Ohio 45217

Pace Internation
500 7th Ave. South
Kirkland, WA 98033

Pestcon Systems
5511 Capital Center Dr. Ste 302
Raleigh, NC 27606-3365

Phelps Dodge Refining Corp.
2600 N. Central
Phoenix, AZ 85002

Phillips Petroleum
Bartlesville, OK 74004

Philom Bios Inc.
15 Innovation Blvd.
Saskatoon SK S7N2X8 Canada

Plant Health Technologies
P.O. Box 15057
Boise, ID 83715

Plant Products Corp.
5051 41st. St.
Vero Beach, FL 32960

Platte Chemical Co.
P.O. Box 667
Greeley, CO 80632

Prentiss Drug &
Chemical Co., Inc.
21 Vernon St. CB 2000
Floral Park, NY 11001

Ralston Purina Company
Checkerboard Square
St. Louis, MO 63188

Regal Chemical Co.
P.O. Box 900
Alpharetta, GA 30239

Rentokil Laboratories
Felcourt, East Grinstead
Sussex, England

Research Group Technologies
101 N. Wilmote Rd. #600
Tucson, AZ 85711-3335

Rhone Poulenc
P.O. Box 12014
2 TW Alexander Dr.
Research Triangle Park, NC 27709

Rhone Poulenc Agrochemie
BP 9163 14-20 rue Pierre Baizet
69263 Lyon, France

Richland Corp.
686 Passaic Ave.
West Caldwell, NJ 07006

Rigo Co.
P.O. Box 189
Buckner, KY 40010

Ringer Corp.
9959 Valley View Rd.
Minneapolis, Minn. 55344

Riverdale Chemical Co.
425 W. 194th St.
Glenwood, IL 60425-1584

Rockland Corp.
686 Passaic Ave.
West Caldwell, NJ 07007-0809

Rohm & Haas Company
100 Independence Mall West
Philadelphia, PA 19106

Roussel Bio Corp.
170 Beaver Brook Rd.
Lincolin Park, NJ 07035

Roussel UCLAF
163, Ave. Gambetta
75020 Paris, France

Roussel UCLAF
95 Chestnut Ridge Rd.
Montvale, NJ 07648

Sandoz Agro
1300 E. Touhy Ave.
Des Plaines, IL 60018

Sandoz Agro
Agrochemical Dept.
Basel, Switzerland CH-4002

Sankyo Co., Ltd.
No. 7-12, Ginza, 2-Chome
Chuo-ku
Tokyo 104 Japan

Sapporo Breweries, Ltd.
10-1 Ginza 7-Chome
Chuo-ku,
Tokyo 104 Japan

SARIAF
20124 Milano
Italy

Scentry Inc. (Ecogen)
610 Central Ave.
Billings, MT 59109

Schering AG (Now AgrEvo)
Postfach 650311
D-1000 Berlin 65,
Germany

The Scotts Co.
14111 Scottslawn Rd.
Marysville, OH 43041

SDS Biotech KK
Higashi Shinbashi Bldg.
12-7 Higashi Shinbashi 2-Chome
Minato-ku
Tokyo 105 Japan

SSI Mobley
1909 N. Longview
Kilgore, TX 75662

Sepro Corp.
11550 N. Meridian St. #200
Carmel, IN 46032

Shin-Etsu Chemical Intl.
Asahi Tokai Bldg.
6-1 Ohtemachi 2-chome
Chuydo-ku
Tokyo 100 Japan

Shionogi and Co., Ltd.
1-8 Doshomachi 3-Chome
Osaka, 541 Japan

J.R. Simplot Co.
P.O. Box 198
Lathrop, CA 95330

SKW Trostberg Ag
Postfach 1262
D-83308 Trostberg, Germany

The Solaris Group
P.O. Box 5006
San Ramon, CA 94583

Sostram Corp.
70 Mansell Ct., Ste 230
Roswell, GA 30076

Source Technology Bilologicals
3355 Hiawatha Ave. S.
Minneapolis, MN 55406

Southern Agricultural Insecticide
P.O. Box 218
Palmetto, FL 34220

Sumitomo Corp.
Plant Protection
5-33 Kitahama 4-chome
Chuo-ku Osaka 541
Japan

Sumitomo Chemical Americas
1330 Dillon Heights Ave.
Baltimore, MD 21258

Summit Chemical Co.
7657 Canton Center Dr.
Baltimore, MD 21224

Sureco
310 MLK Jr. Dr.
Fort Valley, GA 31030

Takeda Chemical Industries
12-10 Nihonbashi 2-Chome
Chuo-ku
Tokyo 103 Japan

Tifton Innovation Corp.
P.O. Box 1753
Tifton, GA 31794

Terra International Inc.
P.O. Box 6000
Sioux City, IA 51102-6000

Tokuyama Soda Co. Ltd.
4-5 Nishi-shimbasti
1-Chrome Minato-ku
Tokyo 105 Japan

Tomen Agrochemicals
14-27 Akasaka 2-Chome
Minato-ku,
Tokyo 107 Japan

Tomen Pacific Agro
444 Market St. Ste1060
San Francisco, CA 94111

Tosoh Corporation
1-7-7 Akasaka
Minato-ku,
Tokyo 107 Japan

Trical
P.O. Box 1327
Hollister, CA 95024-1327

Troy Biosciences
2620 N. 37th Dr.
Phoenix, AZ 85009

UBE Ind. Ltd.
UBE Bldg.
311 Higashi-shingawa 2-chome
Shinagawa-ku
Tokyo 140 Japan

UCB Agrochemical Ltd.
Avenue Lowse 326
B-1050 Brussels, Belgium

UCB Chemicals Corp.
5505-A Robin Hood Rd.
Norfolk, VA 23513

Uniroyal Chemical Co.
Crop Protection Division
Middlebury, CT 06749

United Agri Products Inc.
P.O. Box 1286
Greeley, CO 80632

Unocal Chemical Corp.
2929 E. Imperial Hwy.
Brea, CA 92621

US Borax & Chemical Corp.
26877 Tourney Rd.
Valencia, Ca 91380

Valent Corp.
P.O. Box 8025
Walnut Creek, CA 94596-8025

Van Diest Supply
P.O. Box 610
Webster City, IA 50595-0610

Wacker-Chemie Gmbh.
Hans Seidel-Platz 4
D-8000 Munich 83 Germany

WR Grace & Co.
1 Town Center Rd.
Boca Raton, FL 33486-1010

Webb Wright Corp.
P.O. Box 1572
Ft. Myers, FL 33902

Westbridge Agricultural Products
2776 Loker Ave. West
Carsbad, CA 92008

Western Farm Service
P.O. Box 1168
Fresno, CA 93715

Whitmire Research Labs Inc.
3568 Tree Court
St. Louis, MO 63122-6620

Wilbur Ellis Co.
191 W. Shaw Ave.
Suite 107
Fresno, CA 93704-2876

Zeneca Agro-chem. Ltd.
Fernhurst, Haslemere
Surrey GU27 3JE
England

Zeneca Ag Products
1800 Concord Pike
Wilmington, DE 19897

Zoecon Corporation
1300 E. Touhy Ave.
Des Plaines, Ill 60018

NOTES

NOTES

NOTES